Energy Innovations
2015

Also by Emil Morhardt

The Humanities and Climate Change
Energy, Biology, Climate Change
Global Climate Change and Natural Resources 2014
Global Climate Change and Natural Resources 2013
Ecological Consequences of Climate Change 2012
Global Climate Change and Natural Resources 2011
Ecological Consequences of Global Change 2011
Climate Change and Natural Resources 2010
Ecological Consequences of Global Climate Change: Summaries of the 2009
Scientific Literature
Global Climate Change and Natural Resources: Summaries of the 2007–2008
Scientific Literature
Biology of Global Change
Global Climate Change: Summaries of the 2006–2007 Scientific Literature
Research in Natural Resources Management
Clean Green and Read All Over
Research in Ecosystem Services
California Desert Flowers
Cannon and Slinkard Fire Recovery Study: A Photographic Flora

Energy Innovations
2015

J. Emil Morhardt, Editor

CloudRipper Press

Cutting Edge Books

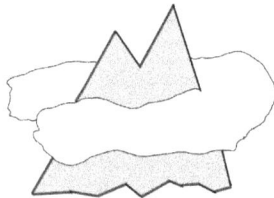

CloudRipper Press
Santa Barbara, California
www.CloudRipperPress.com

Morhardt, J. Emil
    Energy Innovations, 2025
    J. Emil Morhardt, Editor.

ISBN 978-0-9963536-1-8 (paper)

TABLE OF CONTENTS

TABLE OF CONTENTS

FORWARD...21
    by Emil Morhardt

ENERGY EFFICIENCY 23

Coolerado's Air Conditioners Save Energy With HMX Technology .. 23
    by Mariah Valerie Barber

Apple Announces Data Centers run on Renewable Energy............. 24
    by Mariah Valerie Barber

ORNL Creates Low Cost Energy Sensors 25
    by Mariah Valerie Barber

Sabien's Boilers Significantly Reduces Energy Consumption 26
    by Mariah Valerie Barber

Nest: Smart Tech That Pays for Itself.. 27
    by Hannah Brown

Copenhagen, the New City of Lights 29
    by Hannah Brown

A Light That Lasts Until You're Middle Aged 30
    by Hannah Brown

Powerley: Tracking Your Energy Use with an App.......................... 31
    by Hannah Brown

Three Secrets to Energy Efficiency.. 32
    by Jessie Capper

Making Wastewater Treatment Plants' Energy Consumption Net-Zero
.... 34
    by Jessie Capper

LiquiGlide: A Non-Stick Solution to Waste..................................... 35
    by Alex Elder

Energy Efficient Window Shades 36
    by Alex Elder

Improving Energy Efficiency in the Ceramics Industry................... 36
    by Alex Elder

Potential $6 Billion Annual Savings Opportunity via LED Outdoor
Lighting 37
    by Alexander Flores

10 for Tendril... 38
    by Alexander Flores

Surpass with Good Glass...39
   by Alexander Flores

Energy-Friendly Furnace 40
   by Alexander Flores

Stanford's New Energy System......................................41
   by Dylan Goodman

EPA's New Energy Star Home Advisor 42
   by Dylan Goodman

Earthtronics new LED Bulbs 43
   by Dylan Goodman

MIT Energy Initiative 44
   by Dylan Goodman

Staples Recycling and Energy Calculator 45
   by Dylan Goodman

New Water Heater Standards..46
   by Dylan Goodman

SolarWindow Technologies...47
   by Dylan Goodman

Calculating the Benefits: Green Roofs 47
   by Alison Kibe

Using Sound to Detect Air Infiltration Into Buildings 49
   by Alison Kibe

Data Center Energy Consumption 50
   by Briton Lee

Food at What Cost? 51
   by Briton Lee

Purifying the Plastic Process..52
   by Briton Lee

Cooling Buildings by Radiating Heat to Outer Space 53
   by Emil Morhardt

Better Sensors in Buildings to Decrease Energy Consumption 54
   by Niti Nagar

Semitrex Launch Promises to Increase Power Supply Energy
Efficiency, Eliminate Phantom Load.............................56
   by Trevor Smith

Cut Down Energy Loss with Essess's Infrared Technology.............57
   by Abigail Wang

Is Gaming the Future of Saving Energy?.......................58
   by Abigail Wang

Hurting the Environment by Going Green......................59
   by Abigail Wang

ENERGY STORAGE 61

Google Rapidly Strives to Improve Its Battery Technology..............62
   by Mariah Valerie Barber

A Very Special Clay...63
   by Hannah Brown

A Tesla Battery for Your Home?...64
by Nour Bundogji

TREASORES Develops Transparent Electrodes with Superior
Performance...65
by Nour Bundogji

Researchers at Clemson University Develop a Super Battery 66
by Nour Bundogji

An Upcoming Method of Energy Storage.......................................68
by Alex Elder

Ice Energy: As Cool as Energy Storage Gets.................................69
by Alexander Flores

Start Up Tumalow—Reducing Peak Energy and Costs 70
by Alison Kibe

Should You Buy Organic? For Batteries, Yes 71
by Alison Kibe

Everything is Better Deep-Fried..72
by Briton Lee

Threading Energy...74
by Briton Lee

Breathing New Life into Aluminum Batteries 75
by Briton Lee

Energy Storage Under Regulatory Uncertainty..............................76
by Emil Morhardt

Used Electric Vehicle Batteries for Home Energy Storage? 76
by Emil Morhardt

Used EV Batteries to Stabilize the Grid 77
by Emil Morhardt

Beefing up a Wind Turbine with Compressed Air...........................78
by Emil Morhardt

Compressed Air Hybrid Vehicles?...79
by Emil Morhardt

Trackside Flywheel Energy Storage in Light Rail Systems..............80
by Emil Morhardt

Off the Grid, Batteries Not Included 81
by Emil Morhardt

Flywheel Versus Supercapacitor for Running a Small Electric Ferry
....82
by Emil Morhardt

Wearable Supercapacitors: Making Devices More Flexible..............83
by Emil Morhardt

Hybrid Energy Storage for CubeSats 83
by Emil Morhardt

Supercapacitors save Windpower Batteries 84
by Emil Morhardt

Storing Wind Energy on Islands is Risky, Economically.................85
by Emil Morhardt

Energy Stored in the Wire? 85
  by Emil Morhardt

Fast Discharge Batteries: Electric Eels........................... 86
  by Emil Morhardt

Battery Imaging Sheds Light on Future Renewable Energy........... 87
  by Niti Nagar

California's Lack of Storage for Clean Energy 88
  by Shannon O'Neill

California Needs to Invest in Renewable Energy Storage Technology
.... 89
  by Melanie Paty

Lithium-Air Batteries: The Next Battery Revolution? 90
  by Chad Redman

Liquid Air Energy Storage 91
  by Chad Redman

Redox Flow Batteries: For Grid Level Storage 92
  by Chad Redman

Flywheel Kinetic Energy Storage: Energy in Motion 92
  by Chad Redman

Pumped Hydroelectric Storage: Putting Gravity to Work........... 93
  by Chad Redman

IMPROVING THE GRID.. 95

New York Plan to Open a Smart Grid Laboratory 95
  by Mariah Valerie Barber

Microgrid Micromanagement 96
  by Briton Lee

Consequences of the Electric Grid's Inefficiency............... 97
  by Shannon O'Neill

The Future of Electricity: Batteries and the Grid............. 99
  by Ali Siddiqui

Europe's Power Grid to be Increasingly Interconnected: Norway
Plans Deep-Sea Cables to Germany, England 99
  by Trevor Smith

Crossing Borders Through Electricity 100
  by Abigail Wang

NOVEL ENERGY APPLICATIONS 103

ORNL Creates Low Cost Energy Sensors.......................... 103
  by Mariah Valerie Barber

Blue Energy Pilot Plant Opens in the Netherlands.............. 104
  by Mariah Valerie Barber

New Printable Circuits Change the Future of Wearable Tech..... 105
  by Alex Elder

Extending the Range of Wireless Charging...................... 106
  by Briton Lee

Peanuts Packing a Punch 107
  by Briton Lee

Electricity From Low-Level Heat 108
  by Emil Morhardt

Bacteria Electrify Sewage, No Methane Needed........................... 109
  by Emil Morhardt

Microbial Electricity from Cyanide!.. 110
  by Emil Morhardt

Jumping-Droplet Electrostatic Energy Harvesting 111
  by Emil Morhardt

Wearable Energy-Generating Cloth May Replace Batteries........... 111
  by Niti Nagar

Synthetic Diamonds Manipulated to be Sensitive Magnetic-Field
Detectors 112
  by Niti Nagar

SolePower: Solving the Mobile Energy Problem 114
  by Shannon O'Neill

Heat Pumps Placed in Frigid Bodies of Water can Nevertheless Offer
Clean, Cheap Heating 115
  by Trevor Smith

PHOTOVOLTAICS.. 117

China and Kenya Partner to Promote Solar Energy Technology.... 117
  by Mariah Valerie Barber

Inspired by Nature: The Bionic Leaf... 118
  by Hannah Brown

Copenhagen, the New City of Lights 120
  by Hannah Brown

PosiGen is Making Solar Accessible.. 121
  by Hannah Brown

From Sunshine Comes Potable Water.. 122
  by Hannah Brown

iGrenEnergi Develops Technology to Boost Productivity of Solar
Panels.. 123
  by Nour Bundogji

A Fully Transparent Photovoltaic Cell... 125
  by Nour Bundogji

A Self-Powered Video Camera that Can Run Indefinitely without an
External Power Supply... 126
  by Nour Bundogji

Collaboration for Efficient and Reliable Solar Tracking Technology for
Commercial Businesses 127
  by Jessie Capper

Pursuing the Next Frontier for Solar Technology: Northrop Grumman
and Caltech's Space Solar Initiative.. 128
  by Jessie Capper

Solar Power in Space... 130
  by Alex Elder

An Off-Grid Renewable Energy Generator 131
    by Alex Elder

Jim Ayala and Hybrid Social Solutions: Innovation in Distributing Solar ....
    by Liza Farr

Solar Fabric is the Second Generation of Solar Technology 133
    by Liza Farr

Solé Power from MiaSolé..134
    by Alexander Flores

New Solar Farm in China..135
    by Dylan Goodman

Perovskite Solar Cells 136
    by Dylan Goodman

Solar Power Duo: A Perovskite and Silicone Based Semiconductor 137
    by Alison Kibe

Material Architecture: Graphene and Carbon Nanotube Applications for Energy 138
    by Alison Kibe

Expanding the Frontiers of Energy: Pay-as-You-Go Energy 139
    by Alison Kibe

Steps in Solar: The Hydrogen Super Emitters 140
    by Alison Kibe

One Step Closer to Cheaper Solar Power.......................................141
    by Briton Lee

Measuring Impacts of Solar Development on Mojave Desert Plants 142
    by Emil Morhardt

Solar Panels Might Not Help CO_2 Reduction Any Time Soon.........144
    by Emil Morhardt

One of the Nation's Biggest Solar Farm Opens in California 145
    by Shannon O'Neill

Debate over Buying or Leasing Solar Panels in Residential Homes 146
    by Shannon O'Neill

Growing Competitiveness of Solar Energy 147
    by Shannon O'Neill

Community Solar Gardens and Other Shared Solar Power Systems Expected to Rise in 2015 148
    by Melanie Paty

The Cutting Edge of Solar..149
    by Chad Redman

A Spherical Solution to Solar Power..150
    by Chad Redman

The Fight for Home Solar Systems 151
    by Abigail Wang

SOLAR THERMAL... 153

BrightSource's Ivanpah CPS Bird Fatalities Controversy.............. 153
by Mariah Valerie Barber

Crescent Dunes Solves the Solar Farm Bird Death Problem with
Energy Storage.. 154
by Liza Farr

Solar Plant Development Impeded by Tax Credit Reductions 157
by Niti Nagar

HYDRO/TIDAL/WAVE ENERGY 159

Clean Current Utilizes Marine Tidal Turbines to Produce Renewable
Energy 159
by Mariah Valerie Barber

Tidal Lagoon Power Reveals Plans to Build World's First Lagoon
Power Plant.. 160
by Nour Bundogji

Carnegie Wave Energy... 161
by Dylan Goodman

Converting Low-Head Hydropower into Air Pressure/Electricity... 162
by Emil Morhardt

Will Windpower Increase Hydroelectric Environmental Impacts? . 163
by Emil Morhardt

Micro-Hydroelectricity in Building Water-Supply Pipes 164
by Emil Morhardt

Fishing Boat Transformed to Harness Energy from Ocean's Waves
.... 165
by Niti Nagar

Portland Plans to Generate Electricity through Water Pipes......... 166
by Niti Nagar

Difficulty of Harnessing the Ocean's Great Power 167
by Shannon O'Neill

CorPower's Recent Breakthrough in Wave Energy Technology 168
by Melanie Paty

Environmentalists Sue Governmental Agencies in an Effort to Help
Pallid Sturgeon in Montana Rivers 169
by Trevor Smith

Proposed Tidal Lagoon Power Plant Could Provide Enough Energy to
Power all of Wales 170
by Trevor Smith

Making Big Waves in Alternative Energies Energy 171
by Abigail Wang

WIND POWER 175

Kenyan Government Solves Energy Issues with Largest Wind Power
Project in Africa 175
by Jessie Capper

The Benefits of Offshore Wind Energy 176
by Alex Elder

Energy Kites: Airborne Wind Turbines 178
by Alex Elder

Smart Source with Skystream 178
by Alexander Flores

Don't Like Turbines? Try Flying a Kite 179
by Alison Kibe

Beefing up a Wind Turbine with Compressed Air 180
by Emil Morhardt

Mini Piezoelectric Wind Turbine.................................181
by Emil Morhardt

Solar and Wind Producers in Japan, Portugal, Design Floating Power
Plants..182
by Trevor Smith

GEOTHERMAL ENERGY 185

Iceland's Deep Drilling Geothermal Energy Project...............185
by Shannon O'Neill

Renewable Energy Resource Revolution in Nicaragua 186
by Shannon O'Neill

AltaRock's Enhanced Geothermal Systems 187
by Ali Siddiqui

The United Forest Service Hopes to Lease Washington National
Forest for Geothermal Development...............................188
by Trevor Smith

New Fluid Promises to Increase Efficacy of Enhanced Geothermal
Systems 189
by Trevor Smith

NUCLEAR POWER 191

Laser Power Systems 191
by Dylan Goodman

Generating Power from Nuclear Wastes 192
by Alison Kibe

Scientists at the National Ignition Facility Bring Us One Step Closer
to Fusion Power...193
by Niti Nagar

The Key to Unlimited Energy: Nuclear Fusion and its Formidable
Challenges..195
by Chad Redman

Major US Nuclear Energy Provider Fights for Clean Energy Money in
Order to Survive 196
by Trevor Smith

VEHICLES 199

Those Cool Electric Cars..199
by Hannah Brown

Hydrogen Powered Trams Make Their Way in China 200
by Hannah Brown

Solar Impulse, the Solar Airplane, Journeys Around the World ... 202
  by Nour Bundogji

Drone Technology Now Used to Monitor, Preserve, and Restore the Environment 203
  by Jessie Capper

The Future of Car Racing: Electrifying the World of Motorsports .. 204
  by Jessie Capper

"Bio-bus" Runs on Treated Sewage and Food Waste 206
  by Liza Farr

Green Cars are Being Replaced by SUVs: The Need for Cheap 207
  by Liza Farr

Piston Power 208
  by Alexander Flores

Compressed Air Hybrid Vehicles? 209
  by Emil Morhardt

Nissan's Electric Vehicle Fleet's Promising Future in Energy Storage and the Private Sector 210
  by Melanie Paty

H2FC SUPERGEN: UK Hydrogen and Fuel Cell Research at its Finest
.... 212
  by Chad Redman

The Economic Benefit of Strict Auto Emission Regulations 212
  by Chad Redman

China to use Electric Vehicles to Solve Energy Variability Problem
.... 213
  by Ali Siddiqui

BIOFUELS AND SYNFUELS 215

Solazyme Algal Biofuel Production in the United States 215
  by Mariah Valerie Barber

The Zero-Waste Zoo: Using Animal Waste as Fuel 216
  by Alex Elder

Solazyme Seeks Solutions 217
  by Alexander Flores

Algae from Wastewater for Biofuels 218
  by Dylan Goodman

Our Future in Feces: Vehicles Fueled by Biomethane 219
  by Briton Lee

Algae Produce More Biofuel When Starved of Nitrogen, But Why? 220
  by Emil Morhardt

Biomass to Butanol via Engineered Yeast 221
  by Emil Morhardt

Study Shows Flawed Experiments used to Support Policies for "Low-Carbon" Biofuels 223
  by Niti Nagar

How to Make Profits from Garbage 224
  by Niti Nagar

One Step Closer to Renewable Propane 225
  by Niti Nagar

Harnessing Energy from Household Plants 226
  by Niti Nagar

The Success of Incentivizing Renewable Energy Projects from
Landfills 227
  by Shannon O'Neill

The Importance of Wood as a Renewable Energy Resource...........228
  by Shannon O'Neill

Does Biofuel Have a Viable Future? 229
  by Ali Siddiqui

Pee-Power Could Save Lives...230
  by Abigail Wang

Gas-to-Liquids Technology 231
  by Alex Elder

GreatPoint Energy Makes a Great Point 232
  by Alexander Flores

Waste and Fuel with InEnTec 233
  by Alexander Flores

HYDROGEN 235

Toshiba Opens Hydrogen Energy Research Center in Tokyo.........235
  by Mariah Valerie Barber

The Better Way to Produce Hydrogen Fuel 236
  by Nour Bundogji

Hydrogen Hungry Bacteria Bring Artificial Leaf One Step Closer to
Viability...237
  by Liza Farr

Ideal Shift from Gasoline to Hydrogen Fuel Cells....................238
  by Alexander Flores

H2FC SUPERGEN: UK Hydrogen and Fuel Cell Research at its Finest
....240
  by Chad Redman

Feeding Cars With Plants 241
  by Abigail Wang

HYDRAULIC FRACTURING AND CARBON
SEQUESTRATION...243

Why Fracking Works (and Sometimes Doesn't)......................243
  by Emil Morhardt

Why Fracking Might Not Work for as Long as We Would Like.......245
  by Emil Morhardt

Tracking Fracking Fluid with Molecular Tracers 246
  by Emil Morhardt

Instead of Flaring Natural Gas at Fracked Oil Wells, Use it to Treat
Fracking Fluid 247
  by Emil Morhardt

Using Supercritical CO_2 Instead of Water for Fracking 248
  by Emil Morhardt

Heavy Oil Production using Fracking and Microwaves 249
  by Emil Morhardt

Fracking: Fix it or Forget It? Global Gas and Oil Prices Falling 250
  by Emil Morhardt

Fracking in South Texas: Spatial Landscape Impacts 251
  by Emil Morhardt

Try Not to Live Too Close to a Fracked Well 252
  by Emil Morhardt

Biotic Impacts of Fracking 253
  by Emil Morhardt

Methane Emissions in Colorado Exceed EPA Estimates; Fracking?
.... 254
  by Emil Morhardt

Unexpectedly High Methane Concentrations over Pennsylvania Shale Gas Fields Too 255
  by Emil Morhardt

How Long will the Fayetteville Fracking Play Last? 256
  by Emil Morhardt

Iceland's Turning Greenhouse Gases Into Stone 257
  by Hannah Brown

Underground Storage of CO_2: Attempts to Eliminate Carbon Emissions 258
  by Nour Bundogji

ENERGY GOVERNANCE 261

Nigerians Push for Renewable Energy to Solve Power Crisis 261
  by Nour Bundogji

The New European Energy Union Faces Some Critiques 262
  by Nour Bundogji

National Parks or Energy: Kenya's Dilemma 265
  by Jessie Capper

A Call for Holistic and Immediate Action: The World's Peaking Renewable Resources 266
  by Jessie Capper

Can California Continue to Lead in the Renewable Energy Industry?
.... 267
  by Jessie Capper

Psychology's Role in Being Green 269
  by Alex Elder

Hamburg is an Industrial City Reborn with a Renewable Energy Economy 270
  by Liza Farr

Obama Makes Moves to Advance Clean Energy in Place of Dirty Fuel
.... 272
  by Liza Farr

Texas Town Move to 100% Renewable Energy Bodes Well for American GHG Reduction............273
by *Liza Farr*

Energy Star Initiative............275
by *Dylan Goodman*

Starting Small: Shifting into Renewable Energy in South Korea ...276
by *Alison Kibe*

A Greener Apple 277
by *Briton Lee*

Obama's Solar Energy Initiative............278
by *Shannon O'Neill*

Obama and Modi Negotiate Renewable Energy in India............279
by *Melanie Paty*

California's Investment in Clean-Tech is Paying Off............280
by *Melanie Paty*

Effects of Utility Scale Solar Energy on Aquatic Ecosystems in the Southwest 281
by *Melanie Paty*

Clean Energy in India's 2015–2016 Budget Proposal............282
by *Melanie Paty*

Costa Rican Electricity in Q1 2015 has been 100% Renewable.....284
by *Melanie Paty*

Delays in the Desert Renewable Energy Conservation Plan 285
by *Melanie Paty*

The Future of Climate Change in Hilary Clinton's 2016 Campaign 286
by *Melanie Paty*

Hawaii Set on Being the First State to Run on 100% Renewable Energy............287
by *Melanie Paty*

How far have we come?............288
by *Ali Siddiqui*

Apple and Google put their Green Thumbs Down............289
by *Ali Siddiqui*

Costa Rica Using Renewable Energy only for Past 75 Days..........290
by *Ali Siddiqui*

The Battle between Ohio and Solar Energy 291
by *Ali Siddiqui*

Quadrennial Energy Review............292
by *Ali Siddiqui*

Drought-Ridden California Looks to Renewable Energy to Keep the Water Flowing 293
by *Trevor Smith*

Major Energy Leaders Look to Lift United States Embargo on Oil Exports 294
by *Trevor Smith*

New European Energy Union Hopes to Loosen Russia's Grip on Its Energy Market 295
by Trevor Smith

Will Sage Grouses Stop Green Energy Development? 296
by Abigail Wang

Protecting Alaskan Wilderness At What Cost? 298
by Abigail Wang

The Unseen Problems with Nevada's Air Quality 299
by Abigail Wang

A Green Makeover for France 301
by Abigail Wang

Crikey! Australia Could Reach 100% Renewables 302
by Abigail Wang

ENERGY FINANCE AND ECONOMICS 305

Crowdfunding Goes Where No Government Has Gone Before: Caring about the Environment 305
by Hannah Brown

Citi Deploys $100 Billion on Clean Energy 306
by Nour Bundogji

Bloomberg Philanthropies Building Greener Cities 308
by Jessie Capper

Making Solar Energy an Option for All Families 309
by Jessie Capper

Are lower oil prices really a concern for the renewable energy industry? 310
by Jessie Capper

The Cost of Solar Power 311
by Alex Elder

Price Premiums for Homes with Rooftop Solar 312
by Liza Farr

Google Makes New Clean Energy Purchases After Abandoning R&D Efforts 313
by Liza Farr

Rise in Global Renewable Energy Investment Reveals Market Trends 315
by Liza Farr

Colorado Utilities Deploy Renewable Energy Because of Lower Costs 316
by Liza Farr

Potential $8-Billion Green Energy Initiative for Los Angeles 318
by Alexander Flores

Funding for Energy Initiatives in Africa: US-Africa Clean Energy Finance Initiative 320
by Alison Kibe

Crowdsourcing to Eliminate Energy Poverty 321
by Alison Kibe

Changes in Clean Energy Investment in 2014: End of Year Recap 322
by Melanie Paty

Global Consequences of Crude Oil Price Reductions 323
by Ali Siddiqui

Are Energy Credits Enough?...324
by Ali Siddiqui

About the Authors 327
Index..329

FORWARD

by Emil Morhardt

Energy seems, on the one hand, to be the most entrenched of commodities—almost all of our electricity comes from large legacy coal- or gas-fired power plants connected to the grid which is, in turn, seamlessly connected to every wall socket in our homes and businesses; and almost all of our vehicle fuel comes from legacy oil refineries, delivered to service stations in tank trucks. It has been this way for my entire lifetime and probably was for most of our parents and even grandparents' lifetimes as well, and it has hardly changed at all. As a high school student I pumped gas in a Texaco station for several years, equipping me well for the self-service pumps we all use today; the pumps have not visibly changed. As a young man I worked as a consultant at many power plants and oil refineries—they haven't visibly changed either.

On the other hand I now drive a hybrid car stuffed with new technology that I can't even begin to understand, and don't need to stop at gas stations as often as before, and I often drive it through the expansive photovoltaic generating stations, and wind turbine farms of the Antelope Valley, the solar thermal power towers in adjacent Lancaster, and the solar thermal troughs in Kramer Junction, all in the Mojave Desert, on my way up to the Owens valley where I have a house along a creek between two of the many hydroelectric power plants fed by the runoff from the Sierra Nevada Mountains. There is clearly a change in our energy mix in the offing, and global warming, or the fear of it, is driving many of these changes, and the many more that are coming.

Because it seems likely that there is money to be made by reducing fossil fuel use, and energy use in general, there is currently an intense amount of entrepreneurial activity surrounding all aspects of our energy supply and usage. This book is an attempt to make some sense of the overwhelming amount of information about this activity streaming down the web (and in the scientific and engineering journals). It is a result of three months in early 2015 of combing through entrepreneurial websites, news items in the press, and a variety of other sources (all well documented in the book) to see what new and exciting developments are occurring in energy. The result is a fascinating look at the types of changes in our energy mix in the near future.

The text consists of more than 250 vignettes of innovative energy projects, many in their earliest stages, organized by type of energy being considered, then by author.

ENERGY EFFICIENCY

One of the largest sources of "new" energy, and the only source of completely clean energy, is the energy that is never collected or consumed. The obvious approach is just to use less energy period; cut back on driving, wear sweaters instead of turning on the heat, that sort of thing. But the same result can be accomplished with no deprivation by increasing the efficiency of the energy we do use. We are seeing this every day. Compact fluorescent bulbs use much less energy for the same amount of light than do incandescent ones; light emitting diodes (LEDs) use even less. Hybrid gasoline/electric vehicles use less energy for the same or better performance than those with conventional internal combustion engines—the next justification for purchasing a hybrid car is likely to be for its massive acceleration as much as for its savings of energy; all-electric vehicles can outperform both in energy savings and performance, although their range limitations will hold them back until some quick charging technology is invented. Modern household appliances use much less energy than their predecessors with no evident decrease in performance. Modern houses and industrial buildings do the same. This section offers many more examples.

Coolerado's Air Conditioners Save Energy With HMX Technology
by Mariah Valerie Barber

During the 2015 AHR Expo that took place January 26-28, 2015, Coolerado received the "AHR Expo Innovation Award," for their new innovative air-cooling technology. Coolerado, an energy company based in Denver, Colorado, has created air conditioners that use a tenth of the energy used by the most commonly used air conditioners or 90% less energy used by the regular air conditioners. Coolerado utilizes its patented heat and mass exchanger (HMX) and fresh air to create cool air, rather than the chemical refrigerants, compressors, and recycled air used by traditional air conditioners. By using fresh air and HMX technology Coolerado both reduces the amount of energy being used and provides the user of its products with cleaner, healthier air (Coolerado.com). A heat and mass exchanger consists of plastic plates that trap water and keep it to one side and then move heat on another side of the plate. The plastic plates are then stacked on top of each other and positioned to create specific channels through which air is directed. Within the air conditioner two airflows exist: there is a channel for the working air and then a channel for the end product of the cool air to be used. The process of cooling the air, as explained by Coolerado, can be broken down into a five-step process.

First the fresh air enters into the system, then dust and allergens are removed from the incoming air, the working air takes out the heat from the fresh incoming air, the byproduct of heat and moisture is then exhausted from the system, and finally, cool air is then pushed into the building. Slowly Coolerado's air conditioning technology is being commercially used more often. For example, Kellogg's cereal company is currently utilizing Coolerado's technology in its Mexicali, Mexico food processing food plant and has since saved an average of $62,000 annual of the costs to cool its factory. Additionally this processing plant has reduced its CO_2 emissions to 27.37 tons per year from 231.41 tons. In addition to the Kellogg's factory, other companies have opted to use Coolerado's air conditioning technology such as AISO.net (Contracting Business). The next step will be for Coolerado's air coolers to become commonly used to cool individual homes.

Coolerado, Inc. (http://www.coolerado.com/)
Contracting Business. Coolerado Helps Kellogg's Save 88% in Cooling Costs
   (http://contractingbusiness.com/commercial-hvac/coolerado-helps-kellogg-s-
   save-88-cooling-costs)

Apple Announces Data Centers run on Renewable Energy
by Mariah Valerie Barber

On Monday February 23rd, 2015 Apple announced that it is going to be investing $1.9 billion in the construction of two huge data centers in Athenry in County Galway, Ireland and Viborg in central Jutland, Denmark. Lisa Jackson, Apple's vice president of Environmental Initiatives, stated, "We believe that innovation is about leaving the world better than we found it, and that the time for tackling climate change is now" (Apple.com). Apple's announcement to take on climate change will be executed by investing in data centers that will run entirely off of renewable energy. Apple stated that the construction of the two data centers is to be completed and made fully operational by 2017. The centers will be used for the storage of Apple products such as iTunes, Siri, iMessage, Maps and other Apple-specific software programs and products. The two data centers will be operating on 100% renewable energy. Although all the specific details have not yet been released, Apple states that each data center will be 166,000 square meters in size and will be using very advanced technology. The computer systems within the data centers will require a great amount of energy to both power them and cool them down. In order to do so the new centers might be utilizing a combination of solar and wind power, although Apple has not formerly committed to either or both energy sources (North American Wind Power). In addition to being powered by renewable energy, the data centers will also incorporate outdoor environmental designs. The data center in Ireland will incorporate an outdoor education space for school groups and field trips. In addition to an outdoor education space, the Ireland data center will also have an outdoor trail around it. Since the computing centers create so much excess heat the data center in Denmark will be taking that excess heat and redirecting it into the heating systems of the central Jutland area to provide homes with heat. So that Apple can redirect this heat, the data centers will be placed next to Denmark's largest electrical substations which already supply heat to the area. Jackson states, "We're excited to spur green industry growth in Ireland and Denmark and develop energy systems that take advantage of their strong wind resources. Our commitment to environmental responsibility is good for the planet, good for our business and good for the European economy." (Apple.com).

Apple, Inc. (https://www.apple.com/pr/library/2015/02/23Apple-to-Invest-1-7-Billion-in-New-European-Data-Centres.html)
North American Wind Power. (http://www.nawindpower.com/e107_plugins/content/content.php?content.13957)

ORNL Creates Low Cost Energy Sensors
by Mariah Valerie Barber

Oak Ridge National Library, the largest US Department of Energy science research laboratory has created new low-cost wireless sensor technology that can be used to monitor the energy consumed by commercial buildings (Ornl.gov). Currently, buildings consume 40% of all energy being consumed in the United States. Most commercial buildings poorly monitor and control their energy consumption. For example, systems in commercial buildings such as heating, ventilation, air conditioning, and electricity often are under controlled and unmonitored. These new sensors have the potential to reduce the energy consumption of buildings by 20–30% (Physics.org).

The sensors use technology that prints circuits, sensors, antennae, and photovoltaic cells and batteries onto very thin and flexible plastic sheets with adhesive peel-and-stick on the back of it. In addition to being able to be printed and installed in buildings very easily, the sensors are extremely low-cost. The sensors require very little power and are entirely wireless. They are to be stuck with the adhesive glue on the back of them to various walls throughout the building. They monitor for outside air, room temperature, humidity, light level, occupancy, and pollutant. They collect data on each factor that they are monitoring and then send that data a main receiver, which receives data from all the other sensors placed in the building. Since they are wireless they are extremely easy to install and place throughout a building.

These ORNL sensors are extremely cheap, ranging from $1-10 per sensor. Currently, the wireless sensors that are available commercially cost around $150 to $300.

As of now the ORNL's sensors are not available for purchase or commercial use, but the ORNL and the US Department of Energy are in the process of negotiating with developers and international electronic manufactures in order to create an agreement so that these sensors can become widely available commercially (Physics.gov).

Oak Ridge National Laboratory. (http://www.ornl.gov/about-ornl)
Physics.org (http://phys.org/news/2015-03-sensors-yield-energy-efficiency.html)
TAGS: ORNL, Oak Ridge National Laboratory, US Department of Energy, Technology, Data, Energy Consumption, Innovation

Sabien's Boilers Significantly Reduces Energy Consumption
by Mariah Valerie Barber

The products of , a UK-based company that focuses on providing commercial boilers and direct fired hot water heater optimization controls, are drastically reducing the amount of energy consumed across the UK and other parts of the world.

Sabien Technology was founded in 2004. Its two main products are the M2G intelligent boiler load and its M1G Direct Fired Hot Water optimization controls. The two technologies have been able to drastically reduce the amount of energy being used and the amount of CO_2 emissions released.

Sabien Technology's M2G boiler load optimization control fixes a very common problem that both commercial and industrial boilers usually have, which is the tendency for them to run on boiler dry cycling. Boiler dry cycling is when a boiler will fire only to recover the heat that is inevitable lost as it permeates through the boilers' casing. This occurrence is also known as the Standing Losses. The problem with boiler dry cycling is that the boiler will still fire up in attempt to recover the standing losses regardless of whether or not the building, the boiler is heating, actually needs to be heated up further or not. The M2G boiler load optimization control can fix the majority of commercial and industrial boilers by preventing them from running the boiler dry cycling. The M2G control is capable of being placed in most existing boiler control systems, such as Building Management Systems. It will also work with natural gas, oil, and LPG fired boiler systems.

Once installed the M2G has been able to reduce gas consumption and CO_2 emissions by up to 25%. Such reductions can lead to major savings that are visible between the first 6 months to 2 years following the installment.

Sabien Technology's M1G, is similar to the M2G in the way that prevents the automatic cycling from occurring in a heater and the heater from continuing to heat the water unnecessarily. The M1G distinguishes when there is a real need to heat the water.

Sabien Technology (http://www.sabien-tech.co.uk/)

Nest: Smart Tech That Pays for Itself
by Hannah Brown

Google gets into the saving the environment game with their learning thermostat, Nest. For $249, this thermostat begins to make homes into the futuristic abodes like those found in the Disney Channel original movie, Smart House, where a teenage computer savant wins a computerized house for his family (imdb.com). But instead of making perfect smoothies and cleaning your room, Nest creates personalized schedules for the temperature of your house, turns down it's power when it senses that no one is at home, shows you your energy efficiency, and can be controlled remotely from your phone (nest.com). Using Nest, people can now prepare a warmed home when they're returning after a cold outing or ensure that they are not unnecessarily chilling an empty building.

Nest was released a couple of years ago, but its reviews were mixed as people questioned its worth and value. Now Nest isn't just attractive for its hip, on trend tech vibe, it seems to be worth every penny. A new study conducted by Nest analyzed data collected by MyEnergy (an acquisition of Nest) on 1,500 homes' energy use before and after installing the thermostat. The results of this study concluded that the smart thermostat pays for itself in 2 years. In other words, it leads to saving of $131–$145 a year per household by saving users 10% to 12% on heating bills and 15% on cooling bills (wired.com).

Almost more impressive than the findings of this one study is that two more studies that were independent of Nest's funding, conducted by the Energy Trust of Oregon and Vectren, corroborated these results (techcrunch.com).

With Nest, people can not only prepare their homes for a warm welcoming, or save money on their bills, they can also track their energy usage. MyEnergy, which used to be run by Nest's General Manager of Energy Services and is now a subsidiary of Nest itself, tracks users' energy patterns and allows them to understand their own ways. In addition to allowing users to get a deeper understanding of their impact and Nest's capabilities, customers can now contact live Nest Energy Advisors to learn the ways of Nest and how they can save the most.

These studies of the monetary value of Nest are particularly interesting in considering how individuals learn to care about climate change. If they can't relate to the impending, but still seemingly distant, effects of climate change, at least they can be motivated by monetary gains. The short-term goal, in the case of using Nest, benefits in the long-term, as it cuts down energy usage and primes people to be aware of the needs for conservation.

"Smart House" imdb.com

Nest Website https://store.nest.com/product/thermostat/

Alba, Davey. "New Data Shows the Nest Thermostat Can Cut Your Heat Bill by 10 Percent" Wired.com February 2, 2015

Russel, Kyle. "Nest Touts Three Studies Climaining Its Thermostat Pays For Itself In Two Years" Techcrunch.com. February 2, 2015

Tweed, Katherine. "Nest Thermostat Pays for Itself in 2 Years, Studies Find." Greentechmedia.com. February 2, 2015

Copenhagen, the New City of Lights

by Hannah Brown

Copenhagen has a lofty goal: to be the world's first carbon-neutral capital by 2025. Just 10 years away, it is implementing various technological tactics that are appealing for their cost-effectiveness, their ingenuity, and their capacity to save energy.

Thanks to an array of sensors embedded in the light fixtures that collect and feed data into software, using a wireless communication system, Copenhagen has already developed technology to help its citizens and encourage energy efficient transportation. Essentially, the city is using LEDs to create a sensory network that can coordinate a vast number of functions and services, some which have already been implemented.

One way in which Copenhagen is working towards its goal is through the introduction of the "green wave," green LED lights embedded in the ground to help bicyclists, riding in their own lanes, sync with their traffic lights, so that they don't have to stop and providing them the safest routes. They are also using LED streetlights that are motion sensitive and brighten only as vehicles get near, and turn down after they have passed. Additionally, there is a mobile tech side to this movement. For example, truck drivers now can use their smartphones to get updates on when the next light will change and bicyclists' movements are tracked through GPS to monitor traffic, and to give them the right of way if there are more than five cyclists at an intersection.

The city has even more ambitious plans for their remote sensing, including informing the sanitation department when the trash needs to be picked up. It will also assist in easing traffic congestion by specifically timing lights and dealing with the difficulties of weather problems, such as predicting where to salt before a snowstorm.

Using LED lights and sensors is not entirely a new idea. Los Angeles already uses sensors the detect traffic congestion and synchronize signals. Companies such as Cisco Systems and Sensity Systems work internationally, from Bangalore to Barcelona, coordinating these systems. But Copenhagen is using them on an even larger scale, and in conjunction with other technology such as solar powered streetlamps with small wind turbines on top and mobile technology that communicates with government systems to achieve goals, such as prioritizing the passage of bicyclists and buses over cars at intersections. Take the truck drivers as another example, while it may seem trivial that they have an app to time their route, it is actually very smart as stopping and starting the truck costs diesel fuel and money, which adds to noise and air pollution. This smart tech is attractive to city managers, as it will save money and energy, while improving the city's life in general. The LED sensors, and the data they collect, provide a technological perspective on encouraging energy efficient behavior and can be an example for cities across the world to follow (nytimes.com).

Cardwell, Diane. "Copenhagen Lighting the Way to Greener, More Efficient Cities" nytimes.com, Dec. 8, 2014.

A Light That Lasts Until You're Middle Aged
by Hannah Brown

A new light bulb is coming up in the LED market. Not just any old bulb, this one is a specifically long-lasting LED light invented by Jake Dyson, the heir to the famous vacuum cleaner technology empire. While this may sound preposterous, this new LED light technology is meant to last up to 40 years. In fact, as the light has yet to go out, it may last even longer!

This is all even more impressive when in comparison to other light technology. A normal light bulb will last you about 2,000 hours. A CFL bulb, 10,000. LEDs are the cream of the crop coming in with a whopping 50,000 hours. But these are just typical LED lights, the Dyson ones outshine them with at least 180,000 hours of light.

Called the "Ariel" suspension light for the British satellite launched in 1962, the bulb gets its long lifespan from a unique cooling system. It uses the same heat pipe cooling technology that was developed for this famous satellite. Originally, the mechanism siphoned heat away from the satellite's microprocessors to maintain a stable temperature while the satellite was in space. While not a rocket ship this time, the six heat pipes and fins work in tandem to eliminate the heat from the light's core, which lowers the temperature of the bulb, making it consume less energy.

This focus on the high temperature of the bulb is a novel direction in LED bulb manufacturing. Most companies do not worry about this aspect, but it is at the crux of the durability issue. It turns out that the excess heat generated by LEDs wears down the performance and lifespan of the bulb, making it die before its potential is reached. With the cooling system in place, the Dyson light fixture lasts longer than any others.

Not only will this light last long into the future, Dyson has engineered it to be as futuristic as possible. For example, it comes with a ZigBee radio that allows its owners to monitor and adjust its electricity consumption from smartphone or tablet. The light also uses touch-sensitive controls for precise illumination adjustments, and comes in two forms—a ceiling light and a desk lamp.

While these fixtures will probably come at a steep price, their durability means that they will last (almost) a lifetime, and eliminate waste and excess electricity usage in the process.

Wong, Raymond. "Dyson leaps into lighting with LED lights engineered to last 37 years" April 11, 2015. Mashable.com

Prindle, Drew. "James Dyson's son built a super-efficient LED light that will last 40 years" October 7, 2014. Digitaltrends.com

Jake Dyson Light. http://www.dyson.com/lighting/csys.aspx

Powerley: Tracking Your Energy Use with an App

by Hannah Brown

A new app, Powerley, gives people a customized look at their daily energy usage. Available whenever, and wherever you want to check in. Well, wherever you and your smartphone travel together (Michipreneur.com).

Vectorform, in partnership with local utility group DTE Energy, has developed Powerley, which makes this information accessible. It is essentially a mobile energy efficiency platform. The app has caught on–in its first 9 months of existence in the mobile world, it has already been downloaded more than 55,000 times.

The app works in conjunction with smart meters that DTE customers have installed in their homes. It allows customers to see their "daily electric and natural gas usage and receive customized energy-efficiency tips." In such, it is a form of home energy management (Greenmediatech.com).

This customization aspect is a key part of Powerley's draw. The director of the product, Kevin Foreman, argues that when people can see the details, they become more invested in clean energy efficiency.

Customers can use the app for real-time energy insights, to track their energy consumption patterns, set goals for themselves and compare their energy patterns to their neighbors. There's nothing like a good keeping up with the Jones' aspect to make you invested.

In addition to tips and data information, the app gives clients weekly energy challenges. Like a fun game, these points go towards earning an avatar and leveling up. However, unlike other energy apps like Simply Energy that turns the process into a game to entice customers, there are no prizes at the end.

Another distinct feature is the use of a gateway from DTE, called an Energy Bridge. This bridge takes readings from the smart meter, already installed, every three seconds.

Only DTE can see your smart meter data. What they give you is analytics they have run, giving you insights into how you can personally improve your energy usage.

Once people have the energy bridge, they can use the PowerScan function on the app to measure wattage of devices by holding their smartphone up to its power cord, using the magnetometer in the phone to read the magnetic field of wire. Its reads are 95% accurate.

After taking this measurement, people can catalogue how often they use this device, and how much it's costing them on their energy bill. While obviously this won't have a huge impact on the usage of large appliances, which take up the most energy, it is getting people excited and ready for disaggregation.

The tagline on Powerley's website makes their mission clear: "powering customer engagement and savings through a revolutionary energy efficiency channel." (http://www.powerley.com)

Less practical, and more like a toy for showing off energy facts, the app is meant to educate people on energy, and their usage. People seem to be digging it, as the app has a 65% retention rate and, on average, saves people 8.3% in energy consumption.

Tweed, Katherine. "After Designing for Disney, Detroit's Vectorform Moves Into Efficiency Apps." April 20, 2015. Greenmediatech.com

Powerly. http://www.powerley.com

Lewan, Amanda. "Vectorform and DTE Launch Powerley to Track Energy Usage" March 8, 2015. Michipreneur.com

Three Secrets to Energy Efficiency
by Jessie Capper

Many people know the three "R"s of sustainability: Reduce, Reuse, And Recycle. Although this is taught fairly early in one's education, the current climate of our planet demonstrates that it is not engrained in, nor maintaining an influential impact on, people's everyday decisions. OPOWER is a company committed to establishing a cloud-based platform that creates a stimulating, educating, and engaging experience to put consumers on a path towards energy efficiency (OPOWER). Although companies like Honeywell, Cisco, Google, and

Microsoft are developing products and technology to detect energy usage in homes, these high-cost and high-tech solutions have proven ineffective. OPOWER has successfully expanded the use of energy efficiency alternatives using three fundamental approaches: behavioral science, analytics, and technology (Laskey and Kavazovic 2011).

Through applying these techniques, OPOWER saved roughly 400 GWh of energy in 2011; this is approximately equal to one-third of the United States' solar capacity. OPOWER recognized that one of the most effective methods was influencing the behavior of consumers. They realized that people were most inclined to conserve energy when shown data that directly compared their respective household's energy use to that of their fellow neighbors. Along with this competitive incentive, descriptive social norms, social approval, and the use of specialized language compelled consumers to make monumental changes in their energy use. An additional ingredient to this success was OPOWER's impressive use of data and technology. Through a persuasive presentation of personal energy-use data, OPOWER generated targeted suggestions to best suit housing lifestyles. As a result, energy efficiency became increasingly more accessible to consumers.

Although OPOWER's models have proven widely effective, they currently face difficulty with properly expanding. Many companies hope to partner with OPOWER to supply their customers with OPOWER's advanced platform, thereby continuing to spread this new level of consumer engagement with energy efficiency (GreenTechMedia, Sept 9, 2014). One area of concern, however, is that OPOWER must ensure that it is capable of controlling this upsized consumer-base. Due to their highly personalized data analysis, targeted advice structure, and close interaction with their consumer, OPOWER's cloud-based platform is unreliable in providing services equally effective and impactful as currently demonstrated. Furthermore, OPOWER will need to continue to update their system to reflect the ever-evolving technologies pertaining to energy efficiency. As the sector of energy efficiency advances, OPOWER must progress with it to ensure their model remains applicable to all consumers regardless of size, location, financial background, and more.

Lacey, Stephen. "OPOWER Enters Rare Partnership With FirstFuel to Expand Into Commercial Building Efficiency." GreenTechMedia. September 9, 2014. http://www.greentechmedia.com/articles/read/opower-partners-with-firstfuel-to-move-deeper-into-commercial-efficiency

Laskey, A., Kavazovic, O., 2011. OPOWER. XRDS 17, no. 4, 47-51. http://www.opower.com/uploads/library/file/15/xrds_opower.pdf

Making Wastewater Treatment Plants' Energy Consumption Net-Zero

by Jessie Capper

Wastewater treatment plants, sometimes euphemistically known as waste resource facilities, hold potential to be one of the leading industries for on-site renewable energy generation and energy conservation. The City of Gresham Wastewater Treatment Plant in Oregon has been a primary example of this. Through renewable energy production and energy-efficiency investments, the plant's energy management team is in the process of completing the first water resource recovery facility in the Pacific Northwest to achieve net-zero energy consumption (Costello Jan 13 2015). Their hope is that the facility will create all its own energy from onsite renewable power using two co-generation engines fueled by biogas and a solar electric system (Costello Jan 13 2015).

With financial assistance from Energy Trust of Oregon and the Oregon Department of Energy, the energy management team at the plant has been able to make their dreams a reality through reducing the plant's need for electricity from the grid, ultimately generating all its energy needs from renewable sources (Costello Jan 13 2015). The Energy Trust of Oregon is an "independent nonprofit organization serving 1.5 million customers of four investor-owned utilities. Energy Trust invests in cost-effective energy efficiency, helps pay the above-market costs of renewable energy, and transforms markets to higher efficiency products and services" (Energy Trust of Oregon). Energy Trust has supported the City of Gresham over the past ten years with technical assistance, product development, and cash incentives to lower the energy-efficiency project costs and the high costs of the two co-generation systems and solar electric system at the wastewater plant. This has been a mutually beneficial partnerships as the energy managers at the plant have been able to achieve their goal, while the Energy Trust continues to learn how to best support other municipalities and project developers; this, in turn, encourages more renewable energy projects in the area.

Treatment facilities, like the Gresham Wastewater Treatment Plant, are energy intensive, so it has seemed unlikely that they could become completely net-zero in their energy consumption. The Gresham plant requires approximately 5,550

Megawatts hours (MWh) of electricity per year, but the solar and biogas fueled systems at the plant are capable of meeting these energy needs. The co-generation from biogas will produce about 5,100 MWh per year while the solar array will generate the additional 450 MWh of renewable electricity (Costello Jan 13, 2015). The success of the Gresham plant's renewable energy production has helped keep ratepayer costs minimal while also increasing the plant's environmental sustainability.

Costello, Maria. "Turning Waste into Energy in Oregon: City of Gresham Wastewater Treatment Plant." Renewable Energy World. January 13, 2015. Accessed April 15, 2015. http://www.renewableenergyworld.com/rea/blog/post/2015/01/turning-waste-into-energy-in-oregon-city-of-gresham-wastewater-treatment-plant?page=all
Energy Trust of Oregon (http://energytrust.org/)

LiquiGlide: A Non-Stick Solution to Waste
by Alex Elder

LiquiGlide, a company founded by a professor at the Massachusetts Institute of Technology, has developed a new technology that enables sticky substances to flow easily across the surface of any container. This technology works by coating the container's interior with a specialized lubricating liquid which makes the surface permanently wet and slippery. This technology was initially marketed towards commercial uses such as glue and ketchup bottles as well as paint cans. Applying LiquiGlide's technology to these sorts of containers would greatly reduce the amount of waste involved when remnants of these products are left over, unable to be used by the consumer due to their viscosity. Widespread implementation of this technology could have major environmental payoffs by reducing waste. In a few years, LiquiGlide expects this technology to be ubiquitous.

However, the technology has important implications beyond ketchup bottles; it can also be applied at a larger-scale industrial level and potentially allow more efficient methods of pumping crude oil. LiquiGlide has already demonstrated that its technology works very well with this substance. Incorporating this technology into the oil and gas industry would greatly decrease the amount of viscous oil lost in oil drums, pipelines, and other surfaces that oil clings to. A system involving LiquiGlide technology would reduce waste in addition to increasing efficiency and reducing the number of cleaning cycles necessary.

Chang, K. (2015). "With New Nonstick Coating, the Wait, and Waste, Is Over." The New York Times.

LiquiGlide Industries: www.liquiglide.com

Energy Efficient Window Shades
by Alex Elder

Many people believe that in order to be more energy efficient, they must make big lifestyle changes. However, energy efficiency can be improved by making simple changes around the house. This idea inspired the New Visual Media Group to create an innovative yet simple technology to help homeowners improve their everyday energy efficiency. The company, founded in Eatontown, New Jersey, has developed a new take on the traditional window shade. They have created a super-thin sheet of polymer film that can be installed inside insulating glazing units (IGUs). IGUs are essentially double-paned windows which improve insulation, and thus the energy efficiency, of the home. The window shade is coated with an ink that can block infrared light, drastically reducing the amount of solar heat allowed into the home through the window. In fact, according to the New Visual Media Group these shades have a solar heat gain coefficient of <0.08.

The shades are available in a variety of colors and offer full privacy, so they are very practical as well as efficient. Additionally, these shades only cost about $5 per square foot, making this new technology affordable to residential consumers. The New Visual Media Group hopes to expand their technology on a larger scale, eventually incorporating their window shades into office buildings, airplanes, and even cars.

Barrinau, T. "This New Energy-Efficient Window Technology Shows Potential." The USGlass News Network. March 27, 2015.
http://www.usglassmag.com/2015/03/this-new-energy-efficient-window-technology-shows-potential/
New Visual Media Group, LLC. http://www.newvisualmediagroup.com/

Improving Energy Efficiency in the Ceramics Industry
by Alex Elder

The ceramics industry is a huge consumer of energy within the manufacturing world. Its energy expenses account for the majority of the operating costs, due to the firing process required to fire the ceramic material in kilns. In order to decrease the energy usage of tileware production, a materials technology company named Lucideon has developed a low-energy firing technology for the creation of ceramic materials. The new process works by incorporating electrodes into the materials during the firing process. The electric fields from the electrodes are able to increase the speed of the sintering procedure. This alternate process reduces the temperature of the kiln needed to fire the ceramics from 2,100 °F to 1,500 °F, cutting the energy costs required to produce these materials by about a quarter. Additionally, this process also cuts firing time down from 45 minutes to about 10 minutes. Lucideon's commercial tests revealed that the tiles produced through this method showed the same strength as the tiles made using traditional, energy-intensive methods.

Not only will this technology cut costs for ceramic manufacturing companies and increase their production efficiency, but it will also reduce the carbon footprint and environmental impact of these companies. Additionally, Lucideon claims that this technology could have important applications outside of ceramics manufacturing. For example, the same electric field technology could be applied to the production of orthopedic materials. Currently, this technology is being used commercially to produce 15 x 15 cm ceramic tiles, but Lucideon hopes to scale up the manufacturing in the near future.

King, A. (2015). New technology means pottery can be made with less energy. The Sentinel. http://www.stokesentinel.co.uk/Fired-new-technology-success/story-26254345-detail/story.html

Lu, A. (2015). New Technologies Could Further Boost Ceramic Industry. International Business Times. http://au.ibtimes.com/new-technologies-could-further-boost-ceramic-industry-1443865

The Low Energy Firing Project – Tileware Applications. Lucideon: Insight Creating Advantage. http://www.ceram.com/industries/ceramics/energy-reducing-firing-technology/tileware-applications

Potential $6 Billion Annual Savings Opportunity via LED Outdoor Lighting
by Alexander Flores

The rise of LED lighting has increased over recent years and can be seen almost everywhere these days. From car lights to house lights, from TV panels to Christmas lights, LED lighting is slowly, but surely paving the way to an energy efficient lighting future. A typical LED bulb uses 80% less energy than a standard incandescent light bulb and will last up to 25 years, which is why many are pushing to substitute them to light the ways along streets and highways. LED lights are also directional light sources in which well-designed fixtures can point the light exactly where the light is needed and prevent light from going where it's unwanted. It costs approximately $10 billion worth of electricity to power the 100 million or so outdoor lights across the United States, which is equivalent to the power consumed by 6 million homes. The technology and products behind LED lighting can cut a typical city's outdoor lighting bill by half or more. By shifting to energy-saving LED light bulbs, local governments can cut operating expenses while reducing their carbon footprint. President Obama issued the Presidential Challenge for Advanced Outdoor Lighting, which aims to upgrade 1.5 million light poles and encourage mayors across the country to adopt LED bulbs for their cities. The Department of Energy (DOE) has done plenty of solid work on demonstration projects to validate the effectiveness of outdoor LED lights and to develop procurement guidelines for interested businesses and communities. At the moment, less than 5% of US outdoor lighting fixtures use LED bulbs, which is why the Department of Energy estimates that a complete shift to LED outdoor lights would save more than $6 billion and prevent 40 million metric tons of carbon dioxide emissions per year. We can only hope that more and more people will seize this great opportunity to save money and energy with such highly efficient lighting.

Horowitz, Noah. LED Outdoor Lighting: $6 Billion Annual Savings Opportunity Ripe for the Picking. The Energy Collective. February 10, 2015. (http://theenergycollective.com/nrdcswitchboard/2191381/led-outdoor-lighting-6-billion-annual-savings-opportunity-ripe-picking)

10 for Tendril
by Alexander Flores

The Tendril team has created next generation energy service products in order to effectively personalize customers' energy-saving tactics with an open, cloud-based software known as the Tendril Energy Services Management (ESM) Platform. The software platform provides the infrastructure, analytics, and understanding required to personalize energy usage for customers. Tendril seeks reliance on the ESM platform to reduce peak loads, lower costs, and maintain grid reliability. The device-agnostic platform can also be quite effective for business owners since it can create a demand management solution that will grow and evolve with a business. The Tendril ESM Platform is built using a Service Oriented Architecture (SOA), which allows it to scale and evolve with a customer's needs while allowing energy providers to leverage advanced segmentation and micro-targeting to deliver personalized and actionable communication via mail, computer, or mobile phone. The company's physics-based model uses actual home data instead of regression-based modeling and averages in order to provide real, personalized forecasts. This approach provides detailed physical descriptions so that the Tendril providers can predict how much energy will be saved on a house-by-house basis considering insulation, changing windows, modifying thermostat set point, and other specific alterations. The model intelligently ingests any type of data delivered to the customers personalized insights, enticing people to act with highly relevant tips and promotions. With the Tendril ESM Platform homeowners can integrate utilities with existing portals or built-in house apps, connect directly to smart meters or integrate with utility MDMS systems, or connect to any device provide with cloud-to-cloud integrations. The household platform is designed to handle data at a utility scale of 7 TB of data a month and allows utility data to be ingested at any frequency. Tendril hopes to see its ESM platforms connect millions of homes and distributed energy-related assets across the globe in order to create new business opportunities, consumers, and the highest engagement rates.

Tendril 1 (http://www.tendrilinc.com/how-we-do-it)
Tendril 2 (http://www.tendrilinc.com/how-we-do-it/analytics)
Tendril 3 (http://www.tendrilinc.com/how-we-do-it/personalization)
Tendril 4 (http://www.tendrilinc.com/how-we-do-it/micro-targeting)
Tendril 5 (http://www.tendrilinc.com/how-we-do-it/engagement-channels)
Tendril 6 (http://www.tendrilinc.com/how-we-do-it/integration)
Tendril 7 (http://www.tendrilinc.com/how-we-do-it/security)
Tendril 8 (http://www.tendrilinc.com/solving-your-needs)

Surpass with Good Glass
by Alexander Flores

We've all heard about transition prescription lenses, but what about transition manufactured glass? View Inc., has developed a unique, patented Dynamic Glass that has the ability to tint electronically in response to the current environment in order to preserve a view and ultimately keep one cool. Think of it as an alive and intelligent window under your control. This Dynamic Glass is capable of being programmed to adjust separate areas of a building at different times of the day using set algorithms depending on when the sun moves or weather changes. This special glass easily transitions through four variable tints providing constant unobstructed views without heat or glare. This is made possible by the use of electrochromic coating of standard float glass, essentially making an insulating glass unit. Control options are made available for specific applications using a Balance of System (BOS) hardware. View Inc. also provides customization of the glass depending customer preference. A customer can select a given size of insulating glass units (IGU), choose an air space option from 9.5 mm to 15.9 mm with argon gas for maximum thermal insulating performance, and choose an inboard (center layer between two glass layers) regarding preferences of tint, lamination, thickness, strength, and coating.

Glass makeups, or different glass selections, exhibit different performance solar heat gain coefficients (SHGC) and can be specified to a zone or group of windows that are controlled together. For instance, one can set zones for each orientation (North, South, East, and West) due to their different sun exposures throughout a day or for different rooms' light exposure. View Inc. provides building thermal management that reduces air conditioning and lighting costs annually while providing the greatest energy savings at peak periods when energy costs are high. This is done by the blocking of up to 90% of solar radiation during peak cooling demand periods for an average of 20% less energy used overall. Soon enough there will be a stronger demand for this type of glass.

View 1 (http://viewglass.com/product/overview/)
View 2 (http://viewglass.com/product/specifying-your-dynamic-glass/)
View 3 (http://viewglass.com/product/intelligent-glass/)
View 4 (http://viewglass.com/product/energy-efficiency/)
View 5 (http://viewglass.com/video-gallery/)

Energy-Friendly Furnace
by Alexander Flores

The US Department of Energy's National Renewable Energy Laboratory has developed a game-changing Optical Cavity Furnace that utilizes optics to heat and purify solar cells with unmatched precision while increasing the cells' efficiency. Solar cells are traditionally tested for mechanical strength, oxidized, annealed, purified, diffused, etched, and layered using thermal or rapid-thermal-processing furnaces that use radiant or infrared heat to rapidly increase the temperature of silicon wafers. The Optical Cavity Furnace combines assets of photonics and tightly controlled engineering to maximize efficiency while minimizing heating and cooling costs. The furnace encloses an array of lamps within a highly reflective chamber in order to achieve unprecedented temperature uniformity. This virtually eliminates energy loss by lining the cavity walls with super-insulating, highly reflective ceramics and using a complex optimal geometric design. The cavity's optimal geometric design uses approximately half the energy of a conventional thermal furnace since water absorbs what would be lost energy. The Optimal Cavity Furnace dissipates energy solely on the target rather than the container too. Its different configurations utilize the benefits of optics to screen wafers that are mechanically strong to withstand handling and processing, remove impurities, form junctions, reduce stress, increase electronic properties, and strengthen back-surface fields. Recent improvements of the furnace have increased its overall efficiency from 16% to 20% and made it capable of processing 1,200 wafers an hour due to shorter processing times, which only take a few minutes. The Optimal Cavity Furnace stands at about a quarter to half the cost of a standard thermal furnace using 40% less electrical power during wafer processing while producing higher quality and more efficient solar cells. It seems like this optimal cavity will instill plenty of optimism in some of the largest solar-cell manufacturers soon enough.

National Renewable Energy Laboratory 1
    (http://www.nrel.gov/docs/fy11osti/52216.pdf)
National Renewable Energy Laboratory 2
    (http://www.nrel.gov/news/features/feature_detail.cfm/feature_id=1629)
National Renewable Energy Laboratory 3 (http://www.nrel.gov/)

Stanford's New Energy System
by Dylan Goodman

For the past several years Stanford University has been working on rolling out plans for a new campus-wide energy system. The new system, called the Stanford Energy System Innovations (SESI), is expected to come online in 2016. Functionally, the system will rely heavily on solar power. Stanford University has teamed up with SunPower, a California-based solar panel design company, to produce their new solar power station. Together they plan to build a 68-megawatt solar station composed of over 150,000 SunPower solar panels covering nearly 300 acres. They also plan to install an additional 5-megawatt rooftop solar system along with the larger station. Together they are expected to provide for up to 53% of Stanford's total energy use (Henry).

The other aspect of the project is the installation of an innovative heat recovery system. The system will work by capturing heat released through the cooling system and redistributing it through an advanced pump system. The new system will be able to recover nearly two-thirds of the heat discharged through the current system and will provide for 90% of Stanford's campus heating demands. The new system will require the retrofitting of 155 buildings and the replacement of over 20 miles of underground piping (Henry). Nonetheless, the heat recovery system will act as an effective means of reducing long-term energy usage. According to Joseph Stagner of the Sustainability and Energy Management Department at Stanford, "After analyzing the energy patterns of the campus, we devised this scheme which would allow us to collect all the waste heat and use it to heat the campus, and that turned out to be a lot more efficient, a lot more economical and it reduced greenhouse gases and water use, and so that's why the campus selected this option." Overall, the new SESI project is expected to save Stanford $420 million in energy costs over the next 35 years.

Henry, Karen. "New Energy System Will Save Stanford $420M." Energy Manager Today. N.p., 20 Apr. 2015. Web.

Lapin, Lisa, and Kate Chelsey. "A New Campus Energy System Cuts Stanford's Greenhouse Gas Emissions by 68 Percent." Stanford News.

Vanech, Tristan. "Stanford Energy Systems Innovations Project Started." The Stanford Daily. N.p., 19 Apr. 2015. Web.

EPA's New Energy Star Home Advisor
by Dylan Goodman

In December, 2014 the Environmental Protection Agency (EPA) launched its new Energy Star Home Advisor, an online tool dedicated to improving energy efficiency for American homeowners. The release came as a part of the EPA's Energy Efficiency Action Week, a weeklong event in which regional EPA offices hosted events dedicated to increasing awareness about energy use and potential energy efficient upgrades. The entire initiative is designed specifically for homeowners and encourages a do-it-yourself approach to upgrading your home. According to EPA administrator Gina McCarthy, "As we enter the winter months, homeowners can use our new Energy Star Home Advisor to increase energy efficiency and save money while reducing greenhouse gas emissions that fuel climate change." Their new website, https://www.energystar.gov, allows users to create a custom home energy profile which in turn provides customized feedback to the user. Based on your home's unique profile, the Energy Star tool recommends prioritized projects to best increase your in-home energy efficiency. Projects range from adding insulation to vulnerable areas to replacing an air filter in your heating or air conditioning unit. Additionally, users are able to create their own list of projects to complete. As your home evolves, so does your energy profile; simply apply the changes in the online tool to stay up to date with your potential home improvement projects. Through the energy star tool, homeowners are able to view the [positive] environmental impacts of their improvements. "When homeowners take advantage of this important tool and increase the energy efficiency of their homes, many families will notice savings on energy bills and improvements in the comfort of their homes." (Gina McCarthy, EPA) In addition to customized suggestions, users have access to general energy star tips on how to save money and energy during the winter season. According to the EPA, the average homeowner spends $2,000 on heating and cooling per year, a cost that could be cut 10% by simply adding insulation and sealing leaks where necessary. Through the Energy Star Home Advisor, homeowners are able better able to track and improve energy efficiency in their home.

Colaizzi, Jennifer. "EPA Announces New Energy Star Tool for Homeowners to Save Money, Energy This Winter." United States Environmental Protection Agency. N.p., 08 Dec. 2014.

Earthtronics new LED Bulbs

by Dylan Goodman

Earthtronics, a Michigan based lighting company dedicated to providing innovative, efficient lighting options, has been producing light bulbs for commercial use since 2007. Since their creation, Earthtronics has produced compact fluorescent light (CFL) bulbs ranging from 5 to 65 watts. CFL bulbs were originally designed as a more efficient substitute to incandescent bulbs; although still widely used, there's been a more recent move from compact fluorescent towards increasingly efficient light emitting diode (LED) bulbs. Earthtronics just recently introduced their efficient new 12-watt LED bulbs. Designed for commercial use, the new 12-watt LED bulbs are produced specifically to be able to replace the previous standard, 18-watt CFL bulbs. The new LED bulbs are designed to be 33% more efficient than standard CFL bulbs and fit into standard sockets without modification. The new 12-watt LED bulbs are available in three different color temperatures—3000k, 4000k, and 5000k—and put out 1,100 lumens of light. The LED bulbs can be used in any socket a standard CFL bulb can be used in making the switch an extremely easy process; simply unscrew the old bulb and swap it out for a new one. Unlike standard CFL bulbs, LED bulbs are capable of producing full output with no delay, making them suitable for applications that demand immediate light. Earthtronics 12-watt LED bulb is designed to last for up to 25,000 hours, or just over 11 years if used for 6 hours daily, 2–3 times longer than the standard 18-watt CFL bulb. The new Earthtronics LED bulb is available in both 12-watt (1,100 lumen) and 10-watt (800 lumen) versions. Both the 12-watt and 10-watt version are available in both G24q and GX23 lamp bases and can be purchased online at http://www.earthtronics.com as well as atselect other retailers.

"New Energy Efficient 12 Watt LED Plug In Lamp from EarthTronics Saves Energy and Performs Longer Than CFLs." Business Wire. N.p., 17 Feb. 2015.

MIT Energy Initiative
by Dylan Goodman

The MIT Energy Initiative (MITEI) has recently announced it will be giving away over $1.65 million in grants under its annual seed fund program. The money will go to support early-stage energy related projects on campus. Over the past 8 years MITEI Seed Funding has provided over $17 million to 140 different energy-related research projects spanning MIT's five schools. For 2015, there will be 11 companies each receiving $150,000 in seed funding (Abraham). There are more applicants than can receive funding, so projects are chosen by their potential to contribute to increased energy research. Projects can vary across a wide array of fields ranging from hydraulic fracturing to new battery technologies.

The MITEI Seed Funding program encourages new ideas. According to Robert Armstrong, director of MITEI, "The MIT Energy Initiative's seed fund awards build on our successful track record of support for innovative thinking around key energy challenges. "There is tremendous potential in these innovative early-stage projects. This round of grants includes important collaborative research efforts that seek to address key global energy and climate challenges." Past seed fund award winners have been successful in a number of startup companies. One successful project was the creation of substance called LiquiGlide, a type of lubricating agent with various applications. According to Kripa Varanasi, lead professor in charge of the LiquiGlide team, "MITEI seed funding was among the early funding I received which laid the foundation for basic research that ultimately led to the startup LiquiGlide that I co-founded. These awards can be instrumental in helping to move projects from the research stage into the real world." Seed funding acts as a way of providing innovative startups with financial resources they wouldn't otherwise have and allows for the development of new companies and technologies that may otherwise not exit. MITEI's seed program is made possible through funding from founding and current members.

Abraham, Melissa. "MIT Energy Initiative Awarding $1.65M in Seed Funds." Product Design & Development. N.p., 13 Apr. 2015. Web.

Staples Recycling and Energy Calculator
by Dylan Goodman

For the past 10 years Staples Inc. has been providing a money-back recycling program for used ink and toner cartridges. They recently announced having collected 450 million cartridges throughout the course of the program, averaging over 60 million cartridges per year for the last four years. The program has helped to recycle 234 million pounds of plastic and metals, equal to about 8,000 garbage trucks worth of potential waste (Reardon). The Staples' ink and toner recycling enables customers to return used cartridges in exchange for reward dollars. Each cartridge garners $2 and customers are allowed to return up to 10 cartridges per month. Used cartridges can be recycled at any Staples store, or even shipped in through an online program. According to Mark Buckley, vice president of environmental affairs for Staples, Staples is "committed to reducing waste going to landfills and reusing valuable resources through our recycling services." In addition to ink cartridges, Staples recycles various office electronics. According to Buckley, "Our small business customers have told us they want to be more environmentally friendly, but they are pressed for time. That's why Staples makes it convenient for them to recycle their old technology devices at Staples stores across the country whether or not they purchased those items from us." Since 2007, Staples has recycled 85 million pounds of office electronics (Reardon). Staples' recycling program has helped to significantly reduce unnecessary waste from printers and other electronics.

In addition to their recycling program, Staples has just rolled out a new online energy calculator which allows customers to enter customized information about their home into an online application and compare their home's energy usage to other nearby homes and business. One interesting feature of the energy calculator is its ability to compare your home's appliances and energy use to similar homes, and provide feedback on what appliances are consuming the most.

Reardon, Kaitlyn. "Staples Celebrates Sustainability Milestone of 450 Million Ink and Toner Cartridges Recycled and Launches New Energy Calculator." MarketWatch. N.p., 14 Apr. 2015. Web.

New Water Heater Standards
by Dylan Goodman

On April 16th, 2015, new residential water heater regulations will go into effect under the National Appliance Energy Conservation Act (NAECA). Products manufactured before the new regulations come into effect can still be bought and installed, however products manufactured April 16th will have to abide by new efficiency standards. One of the more significant changes will be an increased size in units themselves. One of the requirements for water heaters is a higher energy factor rating, or EF. An EF rating is determined by the amount of hot water produced per unit of fuel over a typical day, so water heaters with a higher EF rating provide hot water more efficiently. One of the ways this can be done is to add more insulation. While this does not increase the efficiency of the actual heating process, it can make heated water stay hot for longer periods, reducing the energy spent on heating. Although this is an easy way to increase overall efficiency, it comes at a cost to space. Many smaller homes uses under 55 gallon water heaters. This increase in size, typically around 2 inches in both height and diameter, must be taken into account when considering installation. Some of the larger heaters will need to implement new design technologies. One of the growing designs is an electric heat pump, which transfers heat from the surrounding air. These larger models will see more substantial boosts in efficiency than the smaller 55-gallon heaters. Nonetheless, the new standards should increase efficiency on residential by an average of 4%. While this does not seem like much, this is particularly good for homeowners as water heaters can account for up to 20% of a home's energy costs (Farrel). Homeowners planning on replacing their hot water heaters in the near future will want to take space into consideration because most heaters come at an increased size.

"Big Changes Coming for Water Heater Efficiency. Ready?" National Association For Home Builders. N.p., n.d. Web.
<http://www.nahb.org/generic.aspx?genericContentID=236255>.
Farrel, Mary. "New Water Heater Regulations." Consumer Reports. N.p., 05 Mar. 2015. Web.

SolarWindow Technologies
by Dylan Goodman

SolarWindow Technologies, Inc. is a developing energy company working towards producing potentially revolutionary glass solar panels. Their claim is that the cost of installation will be earned back within a single year (SolarWindow Tech, Inc. website). This is a little hard to believe, considering that conventional solar farms general take 5–10 years for payback and take up valuable land. The company claims that "Exponentially out-performing today's solar photovoltaic (PV) systems is made possible when engineers apply electricity-generating SolarWindow™ coatings to glass. These see-through liquids create electricity-generating glass windows, successfully prototyped in the most aesthetically appealing colors in demand by building architects." Rather than having to build solar generating plants, this technology could be applied to pre-existing glass surfaces on city buildings. According to their modeling, SolarWindow™ will be able to generate 50 times the power of conventional rooftop solar panels while still being 15-times as environmentally friendly. While producing, distribution and installation of such a seemingly great product may likely be more eco-friendly, some of the company's claims seems questionable. While they claim to have filed for numerous patents, it is unclear if these are on any of the processes behind the unique technologies they claim to have developed. Unfortunately, there is little information about this on their website.

Their claim is to use dissolved organic polymers, which can then be applied as a coating to glass surfaces. Supposedly this technology would be low-cost and easy to apply. According to their website, SolarWindow™ could be applied to all sides of a building and would be able to generate electricity even in the shade. While the future of current developments is uncertain, if SolarWindow™ Technologies, Inc. is able to go into production, their SolarWindow™ technology could revolutionize the solar energy market.

SolarWindow™. N.p., n.d. Web. <http://solarwindow.com/>.

Calculating the Benefits: Green Roofs
by Alison Kibe

The known benefits of green roofs are nothing new; they can reduce building heating and cooling costs, aid in the remediation of the "heat island effect" often observed in cities, and reduce storm run-off. With this in mind, and perhaps putting aside risks and costs associated with green roofs, should I install one? Green Roofs for Healthy Cities, a non-profit industry association for green roof companies, strives to make that an easier question to answer via its Green Roof Energy Calculator (GREC) (greenbuilding.pdx.edu). Developed by University of Toronto and Portland State University and funded by The US Green Building Council, GREC estimates annual cost and savings estimates for office and residential buildings with just a few clicks of the mouse.

A paper by David Sailor and Brad Bass detailing the development process of the GREC cites the need for better tools for architects, urban developers, and building owners who may have limited or no experience with green roofs. For those familiar with building energy simulation software, GREC can also help architects and developers understand storm runoff and urban heat impacts while considering limitations to soil depth and the roof area of the building. The calculator takes roof square footage, geographical location, soil depth, and plant coverage into account and gives the option to adjust the provided utility rates.

With the availability of such tools, are we likely to see the adoption of more green roofs? The answer is probably not. Like many other green technologies, there are still improvements to be made before green roofs become economically practical. An article written for Scientific American about New York City's program that offers tax abatements to those who install green roofs has not gone quite as planned. The plant of choice for green roof projects was sedum—a hardy plant that is able to survive the harsh conditions found atop tall buildings. However, it has been found that sedum actually absorbs heat rather than reflecting it during warm months. Sedum is also not as efficient at absorbing water as other plants so minimizes potential benefits in that area.

A study by Jaffal *et al.* also shows that we are still learning how green roofs can be most effective. They found that green roofs in hot environments create the greatest benefits, but can cause a higher heating demand at certain points in the year. Insulation, another means of reducing a building's heating and cooling cost, was also found to impact the effectiveness of green roofs, with greater insulation often reducing the heating and cooling impact of the green roofs.

The complex nature of how green roofs operate highlights the need for a tool like GREC for anyone who is not an expert in the field considering a green roof. The model used in the calculator is simplified, but for what it seeks to accomplish could be a useful tool when trying to design a more energy-efficient building.

Jaffal, I., Ouldboukhitine, S., Belarbi, R., 2011. A comprehensive study of the impact of green roofs on building energy performance. Renewable Energy 43, 157-164.

Kraft, Amy. "Why Manhattan's Green Roofs Don't Work – and How to Fix Them." Scientific American. May 17, 2013.
http://www.scientificamerican.com/article/why-manhattans-green-roofs-dont-work-how-to-fix-them/

Sailor, D., Bass, B., 2014. Development and features of the Green Roof Energy Calculator (GREC). Journal of Living Architecture 3, 35-58.

The Green Roof Energy Calculator
http://greenbuilding.pdx.edu/GR_CALC_v2/savings_v2.php

Using Sound to Detect Air Infiltration Into Buildings
by Alison Kibe

Since 2013, the US Office of Energy Efficiency & Renewable Energy (EERE) has been funding research at the Argonne National Laboratory and the Illinois Institute of Technology to develop technology that uses sound to detect and quantify air infiltration in homes and buildings. If you imagine that it is the dead of winter and you're staring at a hole on the side of your home, the chance you decide that something ought to be done about it is probably high. Air infiltration caused by air leaks and poor insulation is a problem that can reduce the building efficiency and as a result raise heating and cooling bills. What many people don't realize is that all of the small gaps in windows, walls, and doors can add up to a sizable opening. Given their small size, they may go unnoticed unless tested for.

Finding points of air infiltration and quantifying their effect currently relies on pressure tests and visual inspections. The most common of these methods is the blower door test. Essentially, this test takes a large fan and measures airflow through a building. The fan equipment itself is expensive and while it can quantify air infiltration, the method cannot pinpoint where air leaks are within the building. Methods like the tracer gas method can find exactly where infiltration occurs, but are not quantitative. The aim of the new system, called the Acoustic Building Infiltration Measurement System (ABIMS), is to create a single measurement system to replace the blower door and tracer gas tests that can be used on buildings of all sizes and at varying stages of construction.

As of 2012, the International Energy Conservation Code (IECC) requires new buildings be subject to pressure tests with the goal of limiting infiltration. ABIMS could make it easier to find how to best limit infiltration and, because it should be easier to use, allow for the creation of better code and higher rates of compliance. Following air infiltration codes will mean less heat loss or infiltration and translate into lower energy usage and greater cost savings. In addition, because reducing infiltration also stabilizes air pressure within the building, HVAC systems that use pressure to work effectively will also be able to work more efficiently.

The idea of creating a better system is compelling for reasons outside of energy use as well. According to the Whole Building Design Guide, a program of the National Institute of Building Sciences, air infiltration is also important to hospitals thinking about the spread of airborne diseases and to buildings areas facing severe air pollution. With this in mind and considering the potential changes in ease of use and cost feasibility, a system like ABIMS could make important and better contributions to building design decisions.

United States Department of Energy, 2011. Air Leakage Guide. Building Energy Codes Program.

United States Office of Energy Efficiency & Renewable Energy, 2014. Acoustic Building Infiltration Measurement System. Energy.Gov, accessed February 7th, 2015. http://energy.gov/eere/buildings/downloads/acoustic-building-infiltration-measurement-system-abims

United States Office of Energy Efficiency & Renewable Energy, 2014. 2014 BTO Peer Review Presentation - Acoustic Building Infiltration Measurement System.

Anis, W., 2014. Air Barrier Systems in Buildings. Whole Building Design Guide. http://www.wbdg.org/resources/airbarriers.php

Data Center Energy Consumption
by Briton Lee

Data centers in the United States are using an increasing amount of energy, needing 34 500 MW power plants to power them all (Thibodeau 2014). In 2013, data centers used a total of 91 billion kilowatt-hours, and they are projected to hit 139 billion kilowatt-hours by 2020, a 53% increase (ibid). These data centers alone are contributing to an emission of 97 million metric tons of carbon dioxide each year, and account for over 2% of the energy used nationwide (Hamilton 2013, Fehrenbacher 2014). The use of data centers in cloud computing is one of the few industrial uses of electricity that continues to grow as its applications advance. In 2014, Greenpeace's audit of the environmental impact of data centers condemned Amazon, labelling it as the dirtiest and least transparent internet infrastructure company around (Fehrenbacher 2014). How Amazon chooses what types of energy to source its data centers is especially scrutinized because Amazon Web Services is the "hands down leader in public cloud computing" and many different companies rely on its services (Fehrenbacher 2014). While Amazon does have two data centers powered by 100% carbon-free energy, it seems less concerned with transitioning to "cleaner" energies than it is with energy efficiency. It seems to subscribe to the prediction that if companies adopted the most efficient practices currently available, there would be higher economic and environmental benefits, with a 40% reduction in energy use and $3.8 billion in savings. Despite Amazon's reputation of being the "dirtiest", Amazon is working with to construct a 150 MW wind farm in Benton County, Indiana (Tsao 2015). It is clear that Amazon is still dedicated to investing in renewable energy to power infrastructure in the United States, and is looking for alternative and efficient ways to power their massive data centers.

Thibodeau, Patrick. 2014. "Data centers are the new polluters". Computer World. (http://www.computerworld.com/article/2598562/data-center/data-centers-are-the-new-polluters.html)

Tsao, Rhea. 2015. "Amazon to Support Construction of a 150MW Wind Farm for Datacenters' Power Consumption". Energy Trend. (http://www.energytrend.com/news/20150121-8130.html)

Fehrenbacher, Katie. 2014. "Amazon has one of the dirtiest powered clouds around, says Greenpeace". GigaOm. (https://gigaom.com/2014/04/02/aws-has-one-of-dirtiest-powered-clouds-around-says-greenpeace/)

Hamilton, James. 2013. "Datacenter Renewable Power Done Right". Perspectives. (http://perspectives.mvdirona.com/2013/11/datacenter-renewable-power-done-right/#)

Food at What Cost?

by Briton Lee

In a developed society such as in the US, there are many things that we take for granted, chief among them being food. The consumer is divorced from how the food reaches the shelves, as well as the labor/energy costs that go into the process. The food industry is heavily energy-consumptive, and while energy consumption per capita may have fallen by 1% from 2002 to 2007, food-related energy use increased about 8% as more energy-intensive technologies were developed to produce food for our increasing population (Schwartz 2011). In fact, about 80% of the increase in annual US energy consumption is food-related. Some of the significant ways energy is consumed in food production include fossil fuels needed to power machines, synthesis of crop fertilizers, and supplying/transferring water

(Biederman 2015). The biggest energy investment, by far, is the replacement of manual labor with mechanized labor; high-tech hen houses and related egg-harvesting techniques increased energy use per egg by 40% according to the FDA (Schwartz 2011). This trend is emulated in the household, with more energy-intensive appliances being used, such as blenders, food processors, second refrigerators, etc. Since the US has a long-distance shipping economy, there are also massive energy investments in transportation. This type of economy is particularly vulnerable to supply chain disruptions. It is puzzling to see a rich and fertile region like the US Midwest, with enormous primary production potential, not being able to feed itself. This long-distance shipping economy further removes us from our food source and accelerates the deterioration of our connection to the land. As agricultural and environmental problems such as soil erosion and decline of water quality continue, fewer people will feel that such issues are important as food continues to arrive at the table unimpeded. Ultimately, we should be cognizant of how large the costs of energy production and transportation are, and how they continue to grow.

Schwartz, Kelly. 2011. "Food for thought: How energy is squandered in food industry". USA Today. (http://usatoday30.usatoday.com/money/industries/energy/2011-05-01-cnbc-us-squanders-energy-in-food-chain_n.htm).

"Anuga FoodTec 2015: Energy Efficiency in the Food and Beverage Industry". Food&Drink Business Europe. (http://www.fdbusiness.com/2015/02/anuga-foodtec-2015-energy-efficiency-in-the-food-and-beverage-industry/).

"Increasing Energy Efficiency in Food Processing Facilities". Minnesota Technical Assistance Program. (http://www.mntap.umn.edu/food/energy.htm).

Biederman, David. 2015. "Fossil Fuels Are The Food Of Food". Center for Industrial Progress. (http://industrialprogress.com/2015/01/27/fossil-fuels-are-the-food-of-food/).

Purifying the Plastic Process
by Briton Lee

We use plastics every day, but we rarely think about what it takes to make them. The process is not as simple and cheap as people might believe. In fact, plastic production in the US is tedious, expensive, and requires a lot of energy. Researchers at the University of Colorado Boulder have developed a way to dramatically curb the amount of energy needed for the purification process.

The most common type of plastic is polyethylene, produced from monomeric, two-carbon ethylene. Ethylene production exceeds that of any other organic compound, and is co-created with ethane molecules that have similar structures and properties. Separating the ethylene from the ethane molecules is necessary to produce plastic, but the similarities between the two molecules are what make the purification process (energy-intensive distillation) difficult and costly. Plastic production in the US using distillation techniques accounts for more than 46 million megawatt-hours, which is about the same amount of energy produced by seven average-sized nuclear power plants.

To cut the energy necessary for the purification process, researchers have looked to methods similar to affinity columns used to separate molecules in chemical and biological research. The researchers at CU-Boulder find that silver ions are efficient in complexing with ethylene but not ethane molecules. These silver ions act like hands and grip these ethylene molecules, holding onto the molecules as the ethane is flushed out of the system. This process effectively isolates the ethylene molecules. The team at CU-Boulder demonstrated that using a substrate containing silver ions to isolate ethylene could slash the amount of energy needed to produce polyethylene, especially when scaled to an industrial level. The material they developed has 13 times more separating power than previous materials.

One of the drawbacks of the new technique is that silver is sensitive to contaminants. A way to produce substrates with pure silver ions cost-effectively is something that must be considered before becoming commercially viable. A solution to this is developing substrates that can package silver ions, protecting them from contaminants, but still allowing the ions to complex with ethylene. It may be a while before this technology finds its way into industrial applications; however, due to the scale that plastics are produced globally, even saving a little bit of energy goes a long way.

Cowan, Matthew. 2015. New technique could slash energy used to produce many plastics. University of Colorado Boulder. (http://www.colorado.edu/news/releases/2015/04/14/new-technique-could-slash-energy-used-produce-many-plastics)

Cooling Buildings by Radiating Heat to Outer Space
by Emil Morhardt

Global warming is occurring because there is a slight imbalance in the amount of sunlight striking the earth over the amount of heat being lost. The only way the earth can shed heat is by radiating it into space, and the problem is that with the current concentrations of greenhouse gases in the atmosphere, earth isn't quite warm enough to radiate enough heat outward to stabilize the temperature. One way to address this imbalance, often considered in proposals for geoengineering, would be to decrease the amount of sunlight captured by the earth, say, for example, by reflecting some of it back into space so it doesn't have a chance to be absorbed. Another way would be to increase the effectiveness of radiating heat into space, but I haven't seen any proposals for the latter.

Stanford University researchers, however, have just figured out how to accomplish both substantial reflection (97%) of solar radiation and an increase in radiation of heat to space in a single device, such that it can passively cool the air in it by at least 5 °C (9 °F), and theoretically by almost 20 °C (36 °F) if protected from convective warming (from the wind or breezes). Their prototype ejects about 40 Watts per square meter, which they figure could be improved to 100 Watts per square meter. This occurs in full sunlight with no energy expenditure whatever, and is the first time a device such as this has been implemented. It won't solve the global warming problem directly, but once commercialized, it could go a long way toward decreasing the electricity consumption of buildings, much of which is needed only for cooling.

The researchers accomplished this by using what they call photonic techniques; they deposited a series of thin layers of hafnium dioxide (but they could have used the much cheaper titanium dioxide) interspersed with layers of silicon dioxide over a thin layer of silver on a silicon wafer. This carefully engineered sandwich reflects almost all the incoming radiation while simultaneously strongly radiating longwave infrared radiation at just the wavelengths that go straight through the atmosphere into space. This being Stanford, I'll be on the lookout for a Silicon Valley startup commercializing the amazing technology in the not-too-distant future.

Raman, A.P., Anoma, M.A., Zhu, L., Rephaeli, E., Fan, S., 2014. Passive radiative cooling below ambient air temperature under direct sunlight. Nature 515, 540-544.

Better Sensors in Buildings to Decrease Energy Consumption
by Niti Nagar

Buildings account for about 40% of the total energy consumption in the United States. Researchers at the Department of Energy's Oak Ridge National Laboratory (ORNL) predict advanced sensors and control have the potential to reduce the energy consumption by 20–30%. To develop low-cost wireless sensors, ORNL researchers are experimenting with additive roll-to-roll manufacturing techniques. Roll-to-roll is a technology still is development that allows electronics components like circuits, sensors, antennae, and photovoltaic cells and batteries to be printed on flexible plastic.

Director of ORNL's Building Technologies Program, Patrick Hughes says, "It is widely accepted that energy-consuming systems such as heating, ventilating, and air conditioning (HVAC) units in buildings are under, or poorly, controlled causing them to waste energy." Providing control systems with additional information such as outside air and room temperature, humidity, light level, occupancy and pollutants, could increase energy efficiency. However, currently collecting these data is cost prohibitive, whether the information is gathered using inexpensive conventional sensors that must be wired, or by using expensive $150–300 per node wireless sensors. ORNL proposes to solve this issue by creating a new wireless sensor that could reduce costs to $1–10 per node by using advanced manufacturing techniques such as additive roll-to-roll manufacturing. The nodes can be installed without wires using a peel-and-stick adhesive backing. The smart sensors would collect and send data to a receiver, which accumulates data from many peel-and-stick nodes. The idea behind the design is the more information received, the better the building's energy management.

Hughes says, "If commercially available at the target price point, there would be endless application possibilities where the installed cost to improve the control of energy-consuming systems would pay for itself through lower utility bills in only a few years." ORNL is currently in negotiations to establish a research and development agreement with a premier international electronics manufacturer to make the low-cost wireless sensors commercially available.

Oak Ridge National Laboratory. "Energy use in building: Innovative, lower cost sensors and controls yield better energy efficiency." ScienceDaily. 27 February 2015. <www.sciencedaily.com/releases/2015/02/150227144825.htm>

Semitrex Launch Promises to Increase Power Supply Energy Efficiency, Eliminate Phantom Load

by Trevor Smith

Laguna Beach-based technology startup Semitrex launched in February 2015, highlighting new power supply technology which aims to dramatically increase energy efficiency (Energy Industry Today 2015). Semitrex's innovation revolves around replacing the complex, multi-part power supply used in devices that require AC/DC conversion, including everything from televisions to washers and dryers, with microchips embedded with power supply circuits. The power supplies that can be replaced by the chips, aptly named Power Supply System on a Chip, currently require more than 50 discrete components from 14 different manufacturers (Semitrex 2015). By streamlining this process, Semitrex was able to fully redesign the way these power supplies work, creating a new chip that promises to increase energy efficiency and all but eliminate the phantom load 'always-on' devices drain from the grid.

The power supply chips produced by Semitrex work fundamentally differently than the power supplies currently in use for devices requiring AC/DC conversion. The conventional system reduces power supplied by the grid using inductive power transformers which use energy and dissipate heat even when the unit they are intended to supply power to is not on. By looking to capacitors in order to reduce voltage for use by devices, Semitrex was able to design a chip which does not require magnetic fields and loses very little energy as heat. (Semitrex 2015)

In addition to being more energy efficient, the chips will reduce the phantom load drained by devices which are always plugged in. This problem is further compounded by devices which require a constant internet connection even in standby mode; last year Forbes estimated $80 billion was wasted to supply phantom power to these devices. (Clancy 2014) Semitrex promises that the increased efficiency of their chips will allow devices to essentially use no power (Semitrex 2015) while idling, dramatically cutting down on the phantom load that costs consumers significant amounts of money and places a burden on the power grid.

Clancy, Heather. "IEA: $80 Billion In Power Wasted By Connected 'Things'". 7/24/2014. http://www.forbes.com/sites/heatherclancy/2014/07/24/iea-80-billion-in-power-wasted-by-connected-things/

Energy Industry Today. "Semitrex Addresses Global Energy Crisis with Game-Changing Technology for More Efficient Power Consumption." 2/24/2015. http://energy.einnews.com/pr_news/251574134/semitrex-addresses-global-energy-crisis-with-game-changing-technology-for-more-efficient-power-consumption

Semitrex. "Semitrex Technologies", 2015. http://www.semitrex.com/semitrex-technologies.html

Cut Down Energy Loss with Essess's Infrared Technology

by Abigail Wang

In an age where most people strive to make our environment greener, it's hard to know exactly what to do to be energy efficient. Essess, a start-up developed at MIT in Cambridge, Massachusetts, hopes to alleviate this issue by working with the United States government and utilities companies to cut down on energy loss with infrared technology.

Co-founded in 2011 by Vinny Olmstead and Sanjay Sarma, a professor in Mechanical Engineering at MIT, Essess deploys cars that have thermal-imaging rooftop rigs that create heat maps of homes and buildings. This technology detects fixable leaks in places like windows, doors, and walls to point out where home and business owners are losing the most energy. The rigs have long-wave infrared and near-infrared radiometric cameras that capture heat signatures. In order to separate buildings from natural surroundings, a LiDAR system, which is technology used to create high-resolution maps, captures 3D images.

Essess provides information, like household and demographic data, to utilities companies to show which households leak the most energy and which owners are the most likely to make fixes. It can also determine heating, ventilation, and air-conditioning (HVAC) efficiency, which is easily 50% of the total energy a household uses. So far Essess has ran several trials with a variety of customers; it worked with the United States Department of Defense in military bases from upstate New York to southern states in order to see if the technology would be applicable in different climates. The technology rolled out onto the commercial market last November, targeted at power utility companies. It is currently in its fourth iteration, reflecting its on-going process to be used in the real world.

Energy efficiency, and the desire to pinpoint the source of energy waste, is a popular niche for several startups. Essess has a fair share of competitors, including FirstFuel Software, which is also based out of Massachusetts and plans to expand business to remotely evaluate commercial buildings' energy waste. If Essess, and other startups like it, is successful in helping reduce energy loss, individuals and companies alike could save billions of dollars.

Chernova, Yuliya. "Heat-Mapping Startup Essess Picks Up $10.8 Million to Scan for Energy Leaks." Wall Street Journal: 20 November 2014. http://blogs.wsj.com/venturecapital/2014/11/20/heat-mapping-startup-essess-picks-up-10-8-million-to-scan-for-energy-leaks/

Essess Company website (2015). http://www.essess.com/

Matheson, Rob. "Thermal-Imaging Quickly Tracks Energy Leaks in Homes and Buildings." SciTechDaily: 5 January 2015. http://scitechdaily.com/thermal-imaging-quickly-tracks-energy-leaks-homes-buildings/

Is Gaming the Future of Saving Energy?
by Abigail Wang

It's common knowledge that a big problem in environmental issues is the need for people to undertake individual, personal energy-saving initiatives. Scientists and environmental activists trying to push people to care more about personal energy decisions may have found an answer to their struggle in gaming. By making mundane things fun, people might be more open to cutting down energy usage.

A few companies have already developed games for both companies and individuals to implement better energy actions. Energy Chickens, created by a group of researches and developers at Pennsylvania State University, is one of the latest apps on the market. The game assigns a chicken to each household appliance and the user is responsible for keeping the chickens healthy by maintaining low energy consumption. Healthy chickens grow and lay eggs, which can be exchanged for market items to customize your chickens. If a user increases his or her energy consumption with particular appliances, the chickens associated with those products will grow sick and not lay eggs.

This method of turning daily activities into a game, and rewarding those who make changes to their lifestyle, is called gamification. Gamification seems to be working remarkably well; a study released by the American Council for an Energy-Efficient Economy found that gamification could encourage individuals to save up to 10% of their energy. If people were simply better at using less electricity, which accounts for 38% of carbon dioxide emissions in the US, the potential for reducing emissions would be huge.

Other non-profits and companies are trying to get whole communities involved in energy saving. Cool Choices, a Wisconsin-based nonprofit, offers sustainability games, and business consulting and training to companies and schools to help staff and students adopt more green practices. Cool Choices creates games for company employees to compete to save as much energy as possible, and often offers cash prizes as an incentive.

Is this incentivized gaming strategy a long-term solution? It's hard to tell right now, but studies have shown people to maintain behavior changes as long as the incentives are meaningful. Individuals aren't likely to maintain good energy-saving habits without some sort of reward.

Verchot, Manon. "Games Help Save Energy." Scientific American: 18 February 2015.
      http://www.scientificamerican.com/article/games-help-save-energy/
Penn State Studio Lab. "Energy Behavior Change."
      http://studiolab.psu.edu/projects/energy-behavior-change
Cool Choices. http://coolchoices.com/

Hurting the Environment by Going Green
by Abigail Wang

When you think of "going green" you probably assume it's the first step to living a more sustainable lifestyle. Going green, however, might not be as environmentally friendly as people may think.

Chris Kennedy, a University of Toronto civil engineer, published a study in Nature Climate Change arguing that the transition from fossil fuels to electric-powered technology isn't always the best way to lower carbon emissions. Knowing where your electricity comes from in order to power 'eco-alternatives' is crucial in determining whether or not you're really opting for a greener option. Eco-alternatives could get their energy from burning oil and coal, meaning they're probably even less green than simply using traditional energy methods.

Kennedy proposed setting a decision-making threshold to help individuals, corporations, and policy-makers determine when it's a good idea to move from fossil fuel technology to electric power. This process is also known as electrification. He developed a formula that measures the amount of greenhouse gas emitted by power generated from sustainable technology and determined that if a technology produces more than 600 tons of greenhouse gas for every gigawatt of energy produced, people are better off using traditional gas.

Electrification isn't the best option for countries if it simply increases carbon emissions. Countries like India, Australia, and China generate most of their electricity using coal, which produces nearly double the amount of carbon emissions of Kennedy's suggested threshold. Incorporating electrified technologies into countries like these could actually increase carbon emissions and quicken climate change.

Governments who struggle to determine a specific and measurable target for switching over to electric power may gladly welcome Kennedy's findings. The engineer's study also encourages individuals to better understand where their electricity is coming from before adopting supposedly eco-friendly technologies. However, specific regions could have different needs and it'll be difficult to simultaneously set international goals and hold these regions accountable.

"Where you live could mean 'greener' alternatives do more harm than good." University of Toronto Faculty of Applied Science & Engineering: 4 March 2015.
http://www.sciencedaily.com/releases/2015/03/150304104354.htm

Taylor, RJ. "Where You Live Could Mean 'Greener' Alternatives Do More Harm Than Good." U of T Engineering News: 4 March 2015.
http://news.engineering.utoronto.ca/where-you-live-could-mean-greener-alternatives-do-more-harm-than-good/

Smith, Marshall. "Can Green Technology Actually Harm the Environment?" IndustryTap: 17 March 2015. http://www.industrytap.com/can-green-technology-actually-harm-environment/27285

ENERGY STORAGE

Wind and sunlight are intermittent, and thus is the energy derived from them. This is a problem for the electric grid; grid operators would like their energy to be "dispatchable" so they can dispatch it to meet demand. Wind and solar are inherently undispatchable—they come and go and ignore the grid operators needs. If only the energy they produce could be stored, even for a while, so that it was not necessary for the grid operators to maintain some type of "spinning reserve", typically a fossil-fuel-fired power plant burning fuel unproductively to keep its steam up and even its steam turbine spinning so that when called on, the turbine can turn the generator.

To some degree this problem is being ameliorated by the installation of the new generation of natural gas-fired turbines that can be up to speed in five minutes, but often the grid cannot wait five minutes to avoid a temporary brownout. Plus, the higher the penetration of intermittent renewables, the greater the problem becomes, so in order to gradually replace fossil fuel power plants with renewables, storage of energy produced by wind and solar will become mandatory.

Since the native output of photovoltaics is electricity, it would be ideal to store the energy in that form without having to convert it to some other form such as chemical potentials in batteries, or hydraulic head in pumped hydroelectric operations. Any time there is a conversion, the round trip inefficiency wastes electricity, and the loss can be quite significant.

Electricity can be stored as electric charges in capacitors, and increasingly "supercapacitors" are being experimented with for this purpose, being not only highly efficient, but charging and discharging rapidly, but they are expensive. If the conversion to some other type of energy is needed, flywheels are also potentially excellent, sharing the rapid charge and discharge and long lifetimes of supercapacitors, if not quite the round trip efficiency.

Lithium-ion batteries would, except for their expense, and slow rate of charge and discharge also be reasonably efficient.

As discussed below, there is about to be an excellent source of cheap ones.

Wind turbines can as easily create compressed air as they can electricity, so one approach, discussed below, is to use the compressed air to drive turbines, thus the inefficiency is only one way rather than having to convert electricity to compressed air for storage then reconvert it to electricity, but that approach too is under development.

The native output of solar thermal power plants is heat, which can be stored temporarily in molten salt, delaying for a while the generation of steam to turn turbines. The entirely feasible delay of a few hours with very little loss of energy meets the needs of many grid operators who need only to store the energy from the heat of the day to the evening when the lights come on.

There are many other clever ideas in the works. More of them are discussed below.

Google Rapidly Strives to Improve Its Battery Technology

by Mariah Valerie Barber

As Google Inc. continues to expand its business into wider industries such as transportation, health care, robotics, communications, and its own devices, its need for more efficient batteries is growing. Since this expansion Google has substantially invested in battery research, led by the Google X research lab and a collaboration with AllCell Technologies LLC, based in Chicago, that may lead to more efficient and longer lasting batteries. As the Wall Street Journal puts it, Google has entered into the "Battery Arms Race," joining companies like Apple, Tesla Motors Inc., and International Business Machines Corp, for their desperate search for better batteries.

Currently, Google is pursuing 20 different projects that rely directly upon discovering extremely energy efficient batteries that are able to last for a long time, such as the Google Glass, and its project to develop electric battery-powered cars, and drones. When trying to launch Google Glass major issues resulted from the products short battery life when playing videos.

New battery technologies are slowly emerging. However, many challenges have arisen in trying to produce them cheaply. For example, researchers at Lawrence Berkeley National Lab have discovered technologies that utilize a solid-state battery that uses thin-films to transmit energy currents through solids instead of liquids. Such discoveries have allowed thinner batteries to be produced, because they do not rely upon the usual liquid battery, which requires more volume. They are also safer because they do not rely upon a flammable liquid. Solid-state batteries, because of the thin films they utilize, can be produced to be very thin, flexible, and bendable.

Google is currently trying to take the solid-state battery technology and create products that may be used for commercial consumer devices. Not only does Google aim to make its batteries more efficient and longer lasting, it hopes to develop battery technology that could allow for such batteries to be wearable and potentially even implantable into the human body. Google Chief Executive Larry Page stated that the life of batteries in cellphones is a "huge issue" that has "real potential to invent new and better experiences." A potential breakthrough in more efficient and longer lasting batteries could come very soon.

Wall Street Journal (http://www.wsj.com/articles/google-gets-into-battery-arms-race-1428694613)

A Very Special Clay
by Hannah Brown

Children grow up making little animals, cities and civilizations out of clay and play-dough. They mold the flexible material into new worlds with ease and joy. What if materials of that same plasticity could be used in other ways? To power the lights that these children use to make their creations by? Or your smart phone, your computer, your home?

While in its first stages of development, researchers at Drexel University are one step closer to making a malleable, and conductive, power source. Called MXene, this material consists of electrodes made up of two-dimensional titanium carbide particles, made from etching aluminum from titanium aluminum carbide. This material is made using lithium fluoride and hydrochloric acid. When introduced to water, it becomes flexible like clay. This means that the material can be shaped and rolled out, as thin as tens of microns thick, to create any shape necessary for the product at hand. Once it dries, after being molded, it is highly conductive (nature.com).

The researchers, led by Yury Gogotsi, believe that the unique "physical properties of the clay...as well as its performance characteristics, seem to make it an exceptionally viable candidate for use in energy storage devices like batteries and supercapacitors." Even though this is novel, it seems to work efficiently. As a capacitor, it can store 900 farads per cubic centimeter, which is higher than other materials and the researchers' first iterations of the product. It also has a long life span, losing none of its capacitance even after more than 10,000 charge/discharge cycles (txchnologist.com).

The efficiency and promise of this technology is intriguing. I am sure that as the researchers continue to develop it further, the applications will grow and it will be adapted for various products.

Txchnologist staff. "Soft Conductive Clay Shows Promise For Future Batteries" txchnologist.com December 3, 2014

Michael Ghidiu, Maria R. Lukatskaya, Meng-Qiang Zhao, Yury Gogotsi & Michel W. Barsoum. "Conductive two-dimensional titanium carbide 'clay' with high volumetric capacitance" nature.com December 4, 2014

Drexel Nanomaterials Group. MXenes. Nano.materials.drexel.edu

A Tesla Battery for Your Home?
by Nour Bundogji

With the emerging energy storage market, Tesla Motors Inc., best known for their Model S all-electric sedan, announces plans to release a lithium-ion battery to power your home. Elon Musk, Chief Executive Officer of Tesla Motors, stated in an earnings conference call, "We are going to unveil the Tesla home battery, a consumer battery that would be for use in people's houses or businesses fairly soon."

What is this battery?

Well, Musk said that it "will be like the Model S pack: something flat, 5 inches off the all wall, mounted, with a beautiful cover, an integrated bi-directional inverter, and plug and play."

Ben Popper from The Verge says it may be similar to the Toyota Mirai, which "uses a hydrogen fuel cell and gives owners the option to remove the battery and use it to supply electrical power to their homes." The Mirai has been reported to power the average home for a week when fully charged. However, what sets Tesla's battery apart from Toyota's is that it would be a stationary pack that "would live in consumers' homes, instead of their cars." Musk said.

Dana Hull and Mark Chediak from Bloomberg claim, "Combining solar panels with large, efficient batteries could allow some homeowners to avoid buying electricity from utilities." Initially, this sounded like a good deal to me. But then, I realized that Tesla hasn't released a price for this battery; considering that their Model S was $70,000 MSRP, it may not be a total bargain.

Nonetheless, this is a release to look forward to. The battery is said to enter production in about six months. However, Tesla is still deciding on a date for unveiling their 'home battery.'

Hull, D. and Chediak, M. Tesla Wants to Build a Battery for Your House. Bloomberg Business. Feb 11, 2015.
http://www.bloomberg.com/news/articles/2015-02-12/tesla-planning-battery-for-emerging-home-energy-storage-market
Popper, B. Elon Musk says Tesla will unveil a new kind of battery to power your home . The Verge. Feb 11, 2015.
http://www.theverge.com/2015/2/11/8023443/tesla-home-consumer-battery-elon-musk
Yarow, J. TESLA: In 6 months we will start producing battery packs that can power a home. The Business Insider. Feb. 12, 2015.
http://www.businessinsider.com/tesla-in-6-months-we-will-start-producing-battery-packs-that-can-power-a-home-2015-2

TREASORES Develops Transparent Electrodes with Superior Performance
by Nour Bundogji

TREASORES (Transparent Electrodes for Large Area Large Scale Production of Organic Optoelectronic Device), a European funded project led by Empa researcher Frank Nüesch, aimed to develop better performing optoelectronic devices such as solar cells and LED lighting panels. Recently, scientists and engineers at TREASORES have been focusing on developing widely affordable solar energy by looking at low-cost production technologies. "Flexible organic solar cells have a huge potential in this regard because they require only a minimum amount of materials and can be manufactured in large quantities using roll-to-roll (R2R) processing," tells Nüesch to ScienceDaily.com. R2R processing, much like how newspapers are printed, requires that the entire device, including the transparent electrodes, to be flexible.

As of March 2015, the scientists at TREASORES created prototype flexible solar cell modules along with novel silver-based transparent electrodes that outperform currently used materials. Mechanical testing results show that the cells can be repeatedly flexed to 25 mm whilst maintaining their performance. Such cells have also been shown to have a lifetime greater than 40,000 hours (equivalent to four and a half years).

To put into perspective, pervious electrode technologies suffer from 'waviness' or 'roughness' preventing them from being highly efficient. TREASORES' prototype corrects for this by using a flattening layer to allow defect-free deposition of optoelectronic device stacks. Hence, the researchers developed novel ultra-thin, transparent, silver (Ag) electrodes that are not only cheaper than indium tin oxide electrodes, another widely used electrode on the market today, but also outperforms them.

The thin silver films are sandwiched between two metal oxide (MO) layers, which provide a sheet resistance of 6 Ohms/square with an optical transmission of 85%. MO/Ag/MO electrode stacks allow for the construction of more efficient optoelectronic devices compared to other electrode technologies due to the low peak-to-valley roughness of about 20 nm.

The research demonstrated a record efficiency of 7% and commercially acceptable lifetimes when tested on the field. Using the very same electrode materials, the team achieved 17 lm/W for the production of white light organic LEDs (OLEDs) and more than 20 lm/W for organic light-emitting electrochemical cells (OLECs).

The new low-cost electrode substrates already outperform existing conductive oxide electrodes in many ways. "But we must further improve the resulting device yields from large-scale production by reducing the defect density of the substrates," says Nüesch.

The next step, "is to scale up and improve the most promising technologies identified so far, say, to produce barrier materials and transparent electrodes in larger quantities." In other words, TRESORES plans to print in rolls of more than 100 meters in length.

Empa Swiss Federal Laboratories for Materials Science and Technology. "Towards 'printed' organic solar cells and LEDs." ScienceDaily. ScienceDaily, 18 March 2015. <www.sciencedaily.com/releases/2015/03/150318100820.htm>.

Researchers at Clemson University Develop a Super Battery

by Nour Bundogji

A team of researchers at Clemson University has developed a material that acts as a superhighway for ions, not only making batteries more powerful, but also can change how gaseous fuel is turned into liquid fuel, which can help power plants burn coal and natural gas more efficiently.

To fully comprehend what these researchers have accomplished, it helps to know how batteries and fuel cells convert chemical energy into electricity. As a commentator at ScienceDialy.com put it, "A chemical reaction splits fuel atoms into ions and electrons. The ions go through a substance called an electrolyte while electrons zip around a circuit. When the ions and electrons recombine on the other side of the electrolyte, it creates electrical power."

Today, batteries and fuel cells are limited by how fast the ions can pass through the electrolyte. If the ions were sped up, it would yield a more powerful battery or fuel cell. However, the challenge for engineers is finding the correct mix of electrolyte ingredients that would allow the ions to move as fast as possible.

The research team at Clemson University used ceria doped with gadolinia to develop a highly efficient material. Under a microscope, the material looks like a chessboard with many particles packed closely together. These particles, also known as grains, are gadolinia-doped ceria that allow the ions to pass through grains with ease. However, the researchers noticed that gadolinia tends to accumulate at the boundaries of the tiny grains, ultimately slowing down the ions. Thus, the team added cobalt iron oxide to the mix, which would clean out the accumulated gadolinium along the grain boundaries. Their findings, reported in Nature Communication, demonstrate that the addition of cobalt iron oxide allowed the ions "sail through the electrolyte en route to their rendezvous with the electrons."

This material was also shown to be effective in turning chemical energy into electrical power resulting in more powerful batteries and fuel cells. This was possible since cleaning out the boundaries also allowed for eased movement of oxygen ions, which created pure oxygen. Thus, "the same material that enhances power could also be used to create membrane systems that purify gas mixtures." This is promising since oxygen can now replace steam in the process used to turn fuels into liquid. Pure oxygen could also be used to help burn coal and natural gas.

Brinkman, an associate professor in the Materials Science and Engineering department as Clemson University, said, "I'm proud to be a part of this collaboration. I think we're on the cusp of something potentially world-changing. The ability to control the performance of materials by tuning small interfacial regions represents a huge opportunity in the design of materials for use in energy conversion and storage."

Brinkman explained that the material that researchers used to conduct ions typically works at temperatures near 800 degrees Celsius, so it would be too hot to stick in your pocket. The next step is to work with materials that burn at cooler temperatures.

Clemson University. "New material could boost batteries' power, help power plants." ScienceDaily. ScienceDaily, 10 April 2015. <www.sciencedaily.com/releases/2015/04/150410113603.htm>.

Ye Lin, Shumin Fang, Dong Su, Kyle S Brinkman, Fanglin Chen. Enhancing grain boundary ionic conductivity in mixed ionic–electronic conductors. Nature Communications, 2015; 6: 6824 DOI: 10.1038/ncomms7824

An Upcoming Method of Energy Storage
by Alex Elder

California is currently in the forefront of clean energy production, not only in the United States, but in the world. Although their rising production of wind and solar energy yields many benefits for the state and its residents, it also produces some unprecedented problems. Specifically, the increase in energy generation results in complications for managing the electric grid which has to maintain the balance of energy supply and demand. The strain on the grid has resulted in a technological movement to improve energy storage systems which would help relieve some of the issues associated with increased energy production. Because energy storage systems are designed to store and then quickly release energy onto the grid, they are able to prevent a potential supply imbalance which can sometimes be caused by the sporadic influx of solar and wind energy. Battery-based storage systems in particular can store energy from the grid when electric rates are low and discharge it for use during the day. This kind of system is especially useful for banking solar energy, which can then be used at night or when power from the grid is more expensive at certain times of the day.

One new energy startup is currently at the forefront of the emerging energy storage market. Advanced Microgrid Solutions (AMS) recently won a contract from the utility Southern California Edison Co. to build and maintain large installations of battery-based energy storage systems in commercial buildings throughout Southern California (Yamout, 2014). The startup is tasked with finding the best batteries, the best software, and the best technicians in the field to create these microgrids. AMS currently plans to install the first of its lithium-ion battery systems in groups of commercial buildings in Orange County. This type of system can run from several hundred kilowatts up to one megawatt (Wang, 2015).

Equipping individual buildings with this ultramodern battery storage technology enables the building energy load to be shifted from the electric grid to battery power during peak demand periods. This provides a substantial load reduction for electric utility companies in addition to allowing property owners access to more reliable energy and reduced costs during peak energy hours. The biggest challenge facing AMS and their microgrid project is the need to estimate the optimal size of the battery systems to efficiently meet the projected energy needs of the commercial buildings. Additionally, AMS must successfully manage the cost of using thousands of advanced technology batteries in their microgrids. Due to the vast scale of this unprecedented project, its success and effectiveness will not be fully known until after its completion. Advanced Microgrid Solutions expect their first fleet of energy storage systems to be completed by January 1, 2017 (Wang, 2015).

Wang, Ucilia (2015). An Energy Storage Startup You Need to Know About. GigaOm. https://gigaom.com/2015/02/16/this-is-an-energy-storage-startup-you-need-to-know-about/

Yamout, Manal (2014). Advanced Microgrid Solutions Wins Contract to Build Fleet of Hybrid-Electric Buildings for Southern California Edison Co. http://www.advmicrogrid.com

Ice Energy: As Cool as Energy Storage Gets

by Alexander Flores

Ice Energy, a privately held company based in Santa Barbara, CA, has developed a cost-effective air conditioning utility known as the Ice Bear. Essentially, the Ice Bear's primary function is to convert a power-guzzling air conditioner into a more efficient hybrid that consumes 95% less energy during the peak of the day. This supplementary energy storage unit is compatible with 85% of all commercial air conditioning units and simply stores energy at night when electricity generation is cleaner, more efficient, and less expensive, then delivers it during the day. The Ice Bear works in conjunction with refrigerant-based, 4–20 ton package rooftop systems common to most small or mid-sized commercial buildings. One could simply think of an Ice Bear unit as a battery for an air conditioning system. But just how does it work? An Ice Bear unit consists of a large thermal tank, which operates in an Ice Cooling mode and Ice Charging mode to make ice at night to use for cooling the following day. A typical cycle begins with 450 gallons of water being frozen by a self-contained charging system during the Ice Charge mode in the insulated tank. This occurs by the pumping of refrigerant through a configuration of copper coils within the unit. Once the ice is formed, the condensing unit turns off and cooling energy is stored until needed. As the daytime temperature rises during peak hours, noon to 6 pm, the Ice

Bear unit switches to Ice Cooling mode, which utilizes the ice instead of the air conditioning unit's compressor. This cools the hot refrigerant by slowly melting the ice as it travels through a series of copper coils. The ice-cold refrigerant is then pushed through a modified Ice Energy LiquidDX evaporator coil within the air conditioning unit by a small, highly efficient pump. This Ice Cooling cycle lasts for at least 6 hours and then the building's air conditioning unit takes over at that point while the Ice Bear goes back into Ice Charge mode to start the cycle over again. An average reduction of 12 kilowatts of source equivalent peak demand is delivered by an Ice Bear unit, which shifts 72 kilowatt-hours of on-peak energy to off-peak hours. These units can also be configured to provide utilities with demand response on other nearby electrical loads, which can double or triple its peak-demand reduction capacity. Ideally, a shift to such a utility for commercial buildings' air conditioning units would reduce electric system demand and costs, improve electric system load factor, and improve electric system efficiency and power quality. Who knew a little ice could save so much? It's almost too cool to be true.

Ice Energy 1 (http://www.ice-energy.com/)
Ice Energy 2 (http://www.ice-energy.com/technology/ice-bear-energy-storage-system/)

Start Up Tumalow—Reducing Peak Energy and Costs
by Alison Kibe

How can we save energy, build better energy supply networks, and do it so everyone gains? The energy start up Tumalow hopes it has an answer.

Commercial and industrial buildings use more energy than the average household and energy companies must be able to meet these buildings' demand at any given moment. As a consequence, a higher percentage of infrastructure and operating costs are related to these buildings. Energy suppliers add a demand charge that is typically based on the fifteen minutes of highest energy use to a building's monthly electricity bill (National Grid). For example, if a building uses 30 kW for an hour, 10 kW for the rest of the month, and the demand charge rate is ten dollars per kilowatt, the total demand charge will be $300. Demand charges can be 30–70% of commercial buildings' energy costs (Technology and Development Program, 2000). Tumalow aims to cut this percentage with its peak shaving system (Tumalow.com).

Peak shaving systems use software and back up energy storage to stop a building from purchasing energy from the grid during peak usage times or when the software predicts a peak in energy consumption that would set the demand charge. Tumalow pays for the backup batteries and installation and is paid in turn through a portion of savings incurred through use of the system. The installed batteries can take up excess power from the grid and then provide power back to the grid when needed in exchange for payments (Carter, 2014). Lowered costs and additional revenue streams encourage building managers to install battery backup systems and if the Tumalow network grows, could promote a more stabile electric grid.

Currently, Tumalow is working with a New York high school that has installed the system (SunShot 2014). Beyond this, Tumalow was a winning submission to the SunShot Catalyst Business Innovation Competition through the US Department of Energy, and BigApps NYC. This idea is promising, and could have the potential to address other energy issues such as solar and other renewable source energy storage.

Carter, Michael. "Tumalow energy ingenuity proves big." The Polytechnic. March 2014.
    http://poly.rpi.edu/2014/03/26/tumalaw_energy_ingenuity_proves_big/
National Grid, 2015. Frequently Asked Questions: What is Demand?
    https://www.nationalgridus.com/masselectric/business/rates/5_demand_faq.as
    p
SunShot, 2014. Tumalow: A smart network of energy storage. United States
    Department of Energy. http://catalyst.energy.gov/a/dtd/Tumalow-A-smart-
    network-of-energy-storage/72638-29324
Technology and Development Program, 2000. Saving Money by Understanding Demand
    Charges on Your Electric Bill. United States Forest Service.
http://www.fs.fed.us/t-d/pubs/htmlpubs/htm00712373/
Tumalow.com
http://www.tumalow.com/#about

Should You Buy Organic? For Batteries, Yes
by Alison Kibe

Electronic systems often still take hours to charge and require a certain level of proactivity lest you be caught running out the door with a dying phone. However, that may be poised to change. The use of organic molecules has made it possible charge devices in seconds (StoreDot 2015; Liang *et al.* 2015).

The startup StoreDot is developing batteries that contain nontoxic and more environmentally friendly organic molecules to charge smartphones in under a minute and electric cars in five. The use of certain organic molecules, StoreDot says, gives their batteries special electrical properties and increases the number of charging cycles from 500 to 2,000. Their technology may even be found inside some smartphones by the end of this year (Westaway, 2015).

More recently, Liang *et al.* (2015) published their findings on the ultrafast energy storage capability of an n-type conjugated redox polymer. Liang says the charging speed of these batteries is record setting for an organic material (Kever, 2015). The polymer is made of a naphthalene-bithiophene polymer, traditionally used in transistors and solar cells, with the addition of lithium (Liang *et al.* 2015). The application of lithium as a dopant has altered its electrical properties by increasing the doping capacity of batteries from 0.1 to 2.0 (Liang *et al.* 2015). Because of this, these batteries charge much more quickly than inorganic batteries and increase the number of charging cycles to around 3,000 (Kever 2015).

The drawback to these technologies is their storage capacities, which are still smaller than traditional batteries. Although StoreDot aims to produce batteries for commercial phones that can charge in about a minute, the phone will have to be recharged every 4–5 hours. Still, this could be improved and have interesting implications for the adoption of electric vehicles by overcoming issues with charging en route.

Kever, Jeannie. UH Researchers Discover N-Type Polymer for Fast Organic Battery. University of Houston News. April 2015. http://www.uh.edu/news-events/stories/2015/April/0406OrganicBattery

Liang, Y., *et al.*, 2015. Heavily n-Dopable π-Conjugated Redox Polymers with Ultrafast Energy Storage Capability. Journal of the American Chemical Society, DOI: 10.1021/jacs.5b02290. http://pubs.acs.org/doi/pdf/10.1021/jacs.5b02290

StoreDot: Charge Your World With Groudbreaking Organic Compounds. http://store-dot.com/Solutions.html

Westaway, Luke. StoreDot rapid-charging battery heads to smartphones this year. CNET. March 3 2015. http://www.cnet.com/news/storedot-rapid-charging-battery-heads-to-smartphones-this-year/

Everything is Better Deep-Fried
by Briton Lee

Scientists have been searching for a way to make batteries hold longer charges, on both a commercial and industrial scale. South Korean researchers have made headway in this development, creating a form of 3D graphene "pom-poms" that have a much more efficient energy capacitance than normal graphene.

Graphene can be used as a supercapacitor due to its stability, high conductivity, and large surface area. 3D graphene capacitors are even better because their greater surface area enhances their capacitance. Graphene capacitors are relatively simple, with a carbon-only structure, and versatile enough to incorporate into batteries as electrodes. However, current ways of manufacturing graphene electrodes yield thin films that may stack and aggregate, which decreases surface area and makes the resulting material more difficult to process. These issues have led to the development of graphene foams and aerogels, but these can't be used as electrodes because they're too irregular and not as carbon-dense. Thus, scientists are currently looking to develop ways to create 3D carbon nanostructures for potential use as battery electrodes.

The South Korean material scientists created these "pom-pom" like structures by passing a graphene oxide suspension through an ultrasonic nozzle. This process generates graphene oxide droplets that are sprayed into a hot organic solvent – a process very similar to deep-frying. The 160 °C solvent has a reducing agent that reacts with graphene oxide to form graphene sheets that clump together, creating the 3D structure. Water in the graphene oxide droplets evaporates quickly and is not miscible with the organic solvent, and this flash vaporization is likely responsible for the filamentous, outwardly-radiating structure. Through this process, 5 μm microspheres are precipitated out of solution.

While this is not the first time 3D graphene nanostructures have been developed, this process is much simpler and can feasibly be scaled up to industrial applications. Currently, researchers are working to create silicon-loaded graphene particles because silicon is promising for use as electrodes in lithium-ion batteries but have difficulty maintaining structural integrity due to distortion during charge/discharge cycles. This issue is overcome by encapsulating the silicon in a carbon scaffold, which can be done by spraying the adding silicon to the graphene oxide before spraying it into the hot organic solvent.

Though the carbon "pom-poms" are simple, versatile, and scalable, this development is only one piece of the puzzle, and it may be a while before we see its implementation in consumer electronics, such as smartphones, or industrial applications.

Fingas, Jon. 2015. "Deep-fried graphene may be the key to long-lasting batteries". (http://www.engadget.com/2015/01/18/deep-fried-graphene/)

Patel, Prachi. 2015. "Deep-Frying Graphene Spheres for Energy Storage". (http://cen.acs.org/articles/93/web/2015/01/Deep-Frying-Graphene-Spheres-Energy.html)

Threading Energy

by Briton Lee

At Drexel University, researchers have discovered a way to turn textiles into energy storage devices. More specifically, they've figured out how to turn cotton into a capacitive yarn (which can be used to store energy) by a process called natural fiber welding. This development is paving the way for the burgeoning wearable and flexible electronics industry.

Natural fiber welding was pioneered by researchers at the US Naval Academy, and is a way to embed fibers with functional materials to give the fibers different properties, such as capacitance. The first step of the process involves treating the yarn with molten salt, which causes the polymer chains in the fibers to swell and open up. Immediately afterwards, the expanded polymers are embedded with a functional compound—in this case, activated carbon. There are a variety of materials, like different carbon nanomaterials, which can be embedded in the fiber, but since we ultimately want to weave the fibers into a wearable garment, factors such as skin irritation come into play. Activated carbon is commonly used in water filters and does not pose much of a problem in this regard. After the polymers are embedded with activated carbon, a spool then pulls them through a narrow syringe, which presses the carbon into the fibers. The spool is then washed free of the ionic liquids, solidifying the yarn and trapping the carbon in the fiber. The end result is a fibrous material that retains its original flexibility, but has the capacitive properties of activated carbon. The activated fiber is then twisted with a stainless steel yarn in order to increase the conductivity of the material, allowing the material to charge more easily.

The researchers at Drexel find that cotton gives the best capacitance-to-flexibility ratio, but is weak and frays relatively easily in the weaving process. The authors are currently working on developing stronger, more durable fibers that can stand up to the knitting process while maintaining good capacitive properties.

The improvements in this technology are tied to the development of wearable electronics. In this age, we're looking to integrate flexible electronics into our lives, such as watches, glasses, clothes, phones, and other devices. Many electronics are inflexible, due to the limitations of lithium-ion batteries, which cannot be bent and are relatively heavy. Developing malleable, lightweight, energy storing fibers, sets us on our way to support a flexible electronic future.

Faulstick, Britt. 2015. "Holding Energy By The Threads: Drexel Researchers Spin Cotton Into Capacitive Yarn". (http://drexel.edu/now/archive/2015/March/capacitive-yarn/)

Lee, Y.-H. *et al*. Wearable Textile Battery Rechargeable by Solar Energy. *Nano Lett*. **13**, 5753–5761 (2013

Breathing New Life into Aluminum Batteries
by Briton Lee

Lithium-ion batteries have transformed the modern technological world. They boast massive improvements over previous alkaline rechargeable batteries; lithium-ion batteries have better energy capacity and are stable over many charge cycles. However, while lithium-ion batteries are becoming increasingly efficient, we seem to be approaching a developmental plateau. Accordingly, researchers are looking into alternative materials and chemistry to power batteries of the future.

A team at Stanford University is looking to further develop aluminum batteries to lead the way. Previous aluminum batteries have been disappointing, with their capacity dropping precipitously after a few charge cycles, and completely failing after approximately 100 charge cycles. However, aluminum has its benefits: it's cheap, lightweight, and abundant. Additionally, pure aluminum metal is stable enough to use as electrodes, whereas lithium is volatile and has a tendency to short-circuit or deform over charge cycles.

The researchers have tapped into the power of aluminum and created batteries that recharge quickly and cycle over 7,500 times without capacity loss, about 7.5 times longer than lithium-ion batteries. $AlCl_4^-$ and $Al_2Cl_7^-$ ions are used as the electrolyte solution, and are electron sinks that store and release energy during charge cycles. This battery is also much less dangerous than lithium-ion batteries; Stanford professor Hongjie Dai told CNBC, "Our new battery won't catch fire even if you drill through it." Moreover, the battery's plasticity adds to the potential of flexible electronics that are making their way into mainstream applications.

While it may seem like this battery is ready to go toe-to-toe with lithium-ion batteries in the industry, its mediocre, but stable, charge capacity is preventing it from doing so. Compared to lithium-ion batteries, with a capacity 100 to 200 watts/kg, these aluminum batteries only have a capacity of approximately 40 watts/kg. Further research will improve the capacity and efficiency of these batteries, but even as they stand now there are areas where they can be implemented. These batteries could be used in electrical grids where energy storage capacity may not matter as much. Nevertheless, these batteries hold a strong promise for a future powered by cheaper, safer, and more efficient batteries.

Tarantola, Andrew. 2015. "Stanford's aluminum battery fully charges in just one minute". Engadget. (http://www.engadget.com/2015/04/06/stanfords-battery-charges-in-one-minute/)

Timmer, John. 2015. "Flexible aluminum battery charges fast, stable for over 7,000 cycles". Ars Technica. (http://arstechnica.com/science/2015/04/flexible-aluminum-battery-charges-fast-stable-for-over-7000-cycles/)

Energy Storage Under Regulatory Uncertainty
by Emil Morhardt

A fundamental shortcoming of the US electricity grid is shortage of connected storage: the grid operators must instantaneously provide enough electricity to maintain an acceptable voltage and frequency by ramping generation up or down in real time, mostly using expensive CO_2-releasing electricity from natural gas peaking plants, and if there isn't enough demand to accommodate the electricity coming in from wind and solar, it just goes to waste. Amy L. Stein, writing in the Florida State University Law Review (2014), does a much more detailed job than I did in my brief introduction above of describing the existing energy storage facilities operating on the US grid, including hydroelectric pumped storage, compressed air energy storage, batteries, flywheels, and thermal energy, and the multiple grid services they provide. She then goes on to analyze the regulatory uncertainty that is partially responsible for the lack of grid storage, and makes an attempt at figuring out how to minimize it. This is a comprehensive document and worthy of reading by anyone interested in US energy storage initiatives.

Stein, A., 2014 Reconsidering Regulatory Uncertainty: Making a case for energy storage. Florida State University Law Review 41, 697.

Used Electric Vehicle Batteries for Home Energy Storage?
by Emil Morhardt

Why would you want to store electricity generated at home when you have a perfectly good connection to the grid and the power company buys back all the electricity your solar panels generate at the cost you pay for electricity? You wouldn't. But what if the power company started paying much less for the electricity you are shipping back than you could have purchased it from them—as they have been doing recently in Australia (Muenzel *et al.* 2014)—or decided to charge so much for transmission of electricity back to the grid that there was little point in selling it to them in the first place (as it appears they are contemplating doing in California from the radio spots I hear recently.) Now you might want a battery large enough to prevent any of the electricity you generate getting back to the grid, and ideally meeting all of your routine electricity demands. A cost-effective solution might be about to arrive just in time to offset these likely policy changes; used electric vehicle batteries.

New electric vehicle batteries store something on the order of 24 k Wh—the Nissan Leaf, for example (Lacey *et al.* 2013). Around 2019 many of these will have degraded by 20% and need replacing. According to Lacey *et al.*, they ought to be perfectly usable for non-vehicular energy storage applications even when degraded to 50% of their original capacity, and according to Muenzel *et al.*, all you would need in a typical home is 6–7 k Wh to obviate the need of using utility-provided electricity, so they would work fine in most cases with a payback period of 10 years, about the same as the solar panels feeding them. Thus, combining such a battery pack with existing or new residential photovoltaic installations is likely to make good sense about the time they become available.

By the way, doesn't it seem a little shocking that the amount of energy stored in a Nissan Leaf battery, which is good for something on the order of 40 miles, is enough electricity to run your house for three days? That makes it rather clear that improving gas mileage of cars ought to be far more effective at decreasing CO_2 releases than increasing household energy efficiency.

Lacey, G., Putrus, G., Salim, A., 2013. The use of second life electric vehicle batteries for grid support, EUROCON, 2013 IEEE. IEEE, pp. 1255-1261.
Muenzel, V., de Hoog, J., Mareels, I., Vishwanath, A., Kalyanaraman, S., Gort, A., PV Generation and Demand Mismatch: Evaluating the Potential of Residential Storage. http://www.juliandehoog.com/publications/2015_ISGT_PotentialStorage.pdf

Used EV Batteries to Stabilize the Grid
by Emil Morhardt

If a single used electric vehicle (EV) battery, still functional but with it's storage capacity degraded by 20%, could be used to store energy from photovoltaic panels at home, a bank of them could be used to stabilize the whole grid, according to Gillian Lacey and colleagues (2013) at Northumbria University in the UK. Their particular emphasis is "peak shaving"—supplying enough electricity at times of peak demand that additional fossil-fuel generation can remained turned off. This would also help regulate the line voltage and allow some "upgrade deferral"—putting off investing in needed new or more efficient sources of energy. A particular value of this type of storage is that it would be at the low voltage end of the distribution system, closer to the end user, thus decreasing the potential line losses that would occur if peak power were supplied by regular generation systems, and increasing the life of transformers which are particularly stressed under peak loads. It might also mean that this type of storage would be "distributed"—maybe one battery per residential transformer. Since most large-scale current photovoltaic generation stations generate at low voltage, used EV batteries might be useful there as well.

In grid applications, the authors have assumed only a single charge-discharge cycle per day (matching the daily peak—probably exactly what would occur at home PV installations as well), which would be effective in prolonging the life of the batteries.

Lacey, G., Putrus, G., Salim, A., 2013. The use of second life electric vehicle batteries for grid support, EUROCON, 2013 IEEE. IEEE, pp. 1255-1261.

Beefing up a Wind Turbine with Compressed Air

by Emil Morhardt

If your wind turbine isn't going fast enough to meet the demands of the grid, blow on it a little harder—that's the general idea suggested by Sun *et al.* (2014). The concept is a little like a hybrid electric vehicle; if the internal combustion engine isn't going fast enough, give it a little boost from the electric motor connected to it. Except in this case, it's that if the wind turbine isn't going fast enough, goose it with a little compressed air. You might be envisioning a compressed air nozzle pointed at the turbine blades, but there's a better way: use a motor driven by compressed air to speed up the turbine. One novel aspect to this study is that the device envisioned as a compressed-air motor is something called a scroll expander, or scroll-type air motor, a new type of pneumatic drive, but that doesn't seem to be central to the idea—any suitable air-driven motor should work. The main point is to have it integrated with the wind turbine so that when needed, it can help out in the short term.

To function, it needs a source of compressed air, but where that might come from isn't dealt with in the paper. It could be generated by the same wind turbine using excess electricity to run an air compressor, but maybe there are other sources as well, and I think the authors envision a centralized source of it—rather than one compressor per wind turbine, one compressor per field of turbines, all connected with air hoses, and maybe that one a wind turbine whose native output is compressed air rather than electricity. The authors created a test rig in the lab that simulated all aspect of the system, and conclude that 55% of the energy used to compress the air could be expected to be returned from the boosted wind turbine, which is fairly good for pneumatic motors, and the wind turbine itself was much better at providing steady reliable power with the compressed air device than without.

Sun, H., Luo, X., Wang, J., 2014. Feasibility study of a hybrid wind turbine system–Integration with compressed air energy storage. Applied Energy DOI: 10.1016/j.apenergy.2014.06.083.

Compressed Air Hybrid Vehicles?
by Emil Morhardt

The usual candidate power supplies for the non-fossil-fuel part of hybrid vehicles are chemical batteries, supercapacitors, and flywheels, all powered up using electricity, and generating electricity when their power can usefully replace or supplant the main power source, the internal combustion engine. But these types of electrical storage and the motor/generators they utilize are complex, sophisticated, and expensive, and have barely appeared at all in the developing parts of the world where fossil fuel use is growing fastest. Maybe there is a simpler, cheaper option. One possibility is compressed air energy storage. All you need is a tank (cheap), a reversible compressor (fairly cheap), and a way to link it to the engine. That last part is tricky because the general run of such systems work optimally at a specific pressure, but their performance falls off dramatically as pressures in the tank exceed or fall below optimum as would be expected as the tank is being pressurized or depressurized. The simple solution, according to Brown *et al.* (2014), is to use inexpensive check valves on the tank to prevent over-compression and over-expansion, and an infinitely variable transmission between the compressor and the engine that can operate efficiently at a range of tank pressures. The transmission adjusts by changing the number of thermodynamic cycles of the compressor executes per driveshaft rotation.

Nobody is doing this yet in road vehicles, the authors believe, because the standard approach uses the engine as a compressor and requires the addition of variable valve timing with expensive actuators, rather than the relatively inexpensive addition of a dedicated external compressor and transmission. Although the efficiency of the proof of concept system they cobbled together is not very high—only about 10% of the energy converted to compressed air comes back to drive the engine— the authors figure that if the exhaust heat of the engine were used to keep the air tank hot (which would cost nothing in terms of energy) a round trip efficiency of 47% might be realized at a cost lower than that of battery hybrids, and in a package that would last a long time and could be repaired using mechanical capabilities common in the developing world.

Brown, T., Atluri, V., Schmiedeler, J., 2014. A low-cost hybrid drivetrain concept based on compressed air energy storage. Applied Energy 134, 477-489.

Trackside Flywheel Energy Storage in Light Rail Systems

by Emil Morhardt

Light rail systems, like hybrid electric vehicles, use their electric engines to generate electricity when they are slowing down, a process called regenerative braking. In hybrid electric vehicles, the energy usually gets stored in lithium-ion batteries, which work well because they are comparatively light in weight and not overly bulky. If neither of these were constraints, then flywheels or supercapacitors would be a better choice because they can deliver power faster and they take much longer to wear out. Of the two, flywheels are lighter, less bulky, cost less, and have longer lives according to a study by the UK Rail and Safety Standards board (Kadim 2009). If they are installed alongside the tracks rather than on the trains, weight and bulk are not very important but cost and lifetimes still favor flywheels.

Gee and Dunn (2014) set out to estimate how much energy could be saved in an electric light rail system by using flywheel energy storage alongside the tracks; the result is 21.6% assuming the current mix of regenerative and friction braking, with even more savings if friction braking were reduced further. Additional benefits would include a 30% reduction in substation peak power, thus a smaller investment in substations would be required on new installations. The authors didn't do an economic analysis, but cite another paper that suggests a payback period for installing such a system of only five years. What are they waiting for?

Gee, A., Dunn, R., 2014. Analysis of Trackside Flywheel Energy Storage in Light Rail Systems. IEEE Transactions on Vehicular Technology DOI 10.1109/TVT.2014.2361865

Kadhim, R. 2009. "Energy storage systems for railway applications. Phase 1 Report.," Rail and Safety Standards Board Report, T779, Interfleet Technology Ltd., Derby, UK, Sept. 2009.

Off the Grid, Batteries Not Included
by Emil Morhardt

If there's a need for electricity, but there aren't any power lines nearby, the approach of choice today, in sunny climes, is photovoltaic (PV) panels connected to batteries. But the total amount of electricity that can be stored is then dependent on the number batteries, and if a relatively large amount of storage is needed, this could be prohibitively expensive, heavy, and not very portable. With the advent of hydrogen fuel cells small enough to fit into an automobile and operational at low temperatures, perhaps a fully self-contained electrical generation system could be based on PV panels electrolyzing water to make hydrogen gas, which could then be stored in low-pressure tanks in amounts as large as needed. That's what Cabezas *et al.* (2014) decided to experiment with.

The impetus was that in their country, Argentina, there is much need for off-the-grid electricity in isolated communities, military outposts, and mountain huts. The authors put together the system using commercially available components and their own electrolyzer (in which the electricity from the PV panels splits water into hydrogen and oxygen gas) and fuel cell stack (which recombines the hydrogen and oxygen to produce electricity). The gases were stored at a pressure only slightly higher than atmospheric pressure, ruling out the need for expensive compressors or high-pressure tanks.

The whole system was light in weight, easily transportable, and inexpensive, and seems to be a good candidate for development into a unified module that could be purchased off-the-shelf for off-the-grid needs.

The authors suggest that their system has the advantage of not incurring the energy losses involved in charging and discharging batteries but they did not do an analysis that directly compared the cost or efficiency of their system to a reasonable battery storage alternative.

Cabezas, M.D., Frak, A.E., Sanguinetti, A., Franco, J.I., Fasoli, H.J., 2014. Hydrogen energy vector: Demonstration pilot plant with minimal peripheral equipment. International Journal of Hydrogen Energy.

Flywheel Versus Supercapacitor for Running a Small Electric Ferry

by Emil Morhardt

When we think of all-electric cars, we think lithium-ion batteries because they are lightweight and have a high power density. For ships, light-weight doesn't matter so much, and it turns out there are types of shipping routes that don't need very much energy storage: think ferries, specifically the plug-in ferry Ar Vag Tredan (the "electric boat" in Breton), a zero-emission passenger ferry crossing the Lorient roadstead 56 times a day. When parked between trips it can recharge its supercapacitor more-or-less instantly (that's a main feature of supercapacitors—that and their ability to discharge their power equally quickly to meet any need for power the ship may have.) How would a flywheel energy storage system work compared to the existing supercapacitor? That's the question asked in a new paper by Olivier *et al.* (2014).

The answer is, pretty well. This is a full-on engineering paper stuffed to the gills with calculations, looking at the problem every which way (that's what engineers do), but they are still looking at a basic system: optimizing the flywheel, the electrical machine (they mean motor I bet) and the power converter which converts electricity into flywheel spinning and vice versa. They assume the flywheel will last 20 years, the ship crossing 35 times a day (less than it actually does, it looks like), spending ten minutes motoring across—expending 16 kWh— and five minutes at each end loading, unloading, and recharging.

Olivier, J.-C., Bernard, N., Trieste, S., Mendoza, L., Bourguet, S., 2014. Techno-economic Optimization of Flywheel Storage System in transportation, Symposium de Génie Électrique 2014.

Wearable Supercapacitors: Making Devices More Flexible

by Emil Morhardt

Maybe someday you will be able to recharge your gadgets by plugging them into your jacket, which you charged up in a few seconds from a convenient wall plug or maybe even from WiFi. Yu *et al.* (2014), at the School of Chemistry and Chemical Engineering, Nanjing University, in China, have fabricated experimental sheets of flexible layered conductive and non-conductive materials that they envision as eventually wearable. We all get tired of waiting around for batteries to charge, but supercapacitors charge almost instantly. They don't usually have much energy storage capacity though—you don't get as much energy storage per unit weight or volume as you presently can from batteries—but if they are built into something that you need to carry around with you anyway, that might not be so important. A good example is the plug-in electric boat. Boats don't care much about how large or heavy something is, but they need to be fueled rapidly. So if you could store all the energy you need quickly in your jacket, your battery-powered devices could recharge in your pocket, wherever you are.

Yu *et al.*'s flexible supercapacitors, however, have very good energy storage capacity (50–60 Watt hours per kilogram), as good as the lead-acid batteries in your car, but unlike your car battery, they are quite thin and flexible.

Yu, C., Ma, P., Zhou, X., Wang, A., Qian, T., Wu, S., Chen, Q., 2014. All-solid-state flexible supercapacitors based on highly dispersed polypyrrole nanowire and reduced graphene oxide composites. ACS Applied Materials & Interfaces.

Hybrid Energy Storage for CubeSats

by Emil Morhardt

CubeSats are cool. No, actually very cold, since they're out in space. But they are reproducing like rabbits. There are well over 200 of these little 10 cm X 10 cm X 10 cm cube satellites have been launched into orbit by tucking them into the nooks and crannies in the launch vehicles around much larger satellites. (Some are multiples of cubes, 10 cm X 20 cm, or 30 cm.) They need energy. Until now they have been powered in the main by lithium ion batteries like those in your computer, and charged by the photovoltaic panels that make up a CubeSat's skin. The thing is that these batteries don't work very well when they are cold; the speed of electrochemical reactions, just like those of every other chemical reaction, are modulated by temperature—the colder the slower. The current Li-ion batteries don't work at all below –10°C, yet CubeSats headed for deep space are expected to encounter temperatures of –40°C some of the time. So if you have a CubeSat process that needs power at low temperatures or a short-term burst of power faster than the batteries can provide, you need help.

Supercapacitors, even though they don't hold as much charge per volume as a lithium ion battery, don't much care what the temperature is, plus they charge and discharge much faster than batteries. It makes sense then that some enterprising engineers would combine the two systems to get the benefits of both; the engineers in question are at the CalTech Jet Propulsion Laboratory and at Cal State University, Northridge, nearby. They have just completed a study demonstrating that this hybrid system will solve the problem. The general idea is similar to combining supercapacitors with battery banks to help windpower installations integrate better with the grid but in this case the system is much smaller and in addition to delivering power rapidly without damaging the batteries, it operates at temperatures lower than any wind farm is likely encounter.

Chin, K., Smart, M., Brandon, E., Bolotin, G., Palmer, N., Katz, S., Flynn, J., 2014. Li-ion battery and super-capacitor Hybrid energy system for low temperature SmallSat applications. SSC-14-VII-9, 28th Annual AIAA/USU Conference on Small Satellites.

Supercapacitors save Windpower Batteries
by Emil Morhardt

Windpower, because it is intermittent, works best on the electrical grid if it has some energy-storage facility connected to it. Batteries are the simplest approach, and the low cost of lead-acid batteries makes them good candidates, but they resent being randomly charged and discharged (especially deeply discharged) at the will of the wind, and die prematurely. Enter the supercapacitor; it can be charged by wind turbines much faster than a battery, can deliver its stored energy to the grid much faster as well, and doesn't resent it at all, even being deeply discharged at every cycle. Engineers at Kocaeli University connected a supercapacitor in parallel with a battery so that it would buffer transient current surges, saving the battery to do what it does best. The system worked just like one might expect. There are some graphs in the paper showing just how the current flowed, and I think it is a nice example of an experimental setup to look into these types of hybrid energy storage systems.

Erhn, K., Aktas, A., Ozdemir, E., 2014. Analysis of a Hybrid Energy Storage System Composed from Battery and Ultra-capacitor, 7th International Ege Energy Symposium & Exhibition, June 18-20, 2014, Usak, Turkey

Storing Wind Energy on Islands is Risky, Economically

by Emil Morhardt

Electrical storage is needed to meet demand when solar and wind power become intermittent, particularly in off-grid systems like isolated oceanic islands. Both fast-response and long-term storage is needed; fast response (seconds to minutes) systems to prevent grid overload during short periods of high load (everybody on the island turns on the air conditioners at once), or of clouds passing over sun, or the wind stopping for a while. Longer-term storage (hours to days) can shift the time the power from renewables is produced to when it is really needed, or is at its highest value (more valuable than, say, the diesel generator sets that the island was previously dependent on. But at what cost? Moazeni *et al.* (2014) modeled the expected revenues to be gained in an isolated island environment of intermittent wind, shifting electrical loads, and a pumped storage hydroelectric system station. It turns out that optimizing the amount of storage and using it in the most economically efficient manner is no trivial matter, with the economic returns deviating as much as 29% from what might have been expected by the investors, depending on how the system is managed. What such systems operators apparently clearly need is electrical engineering modelers like these authors to figure it out for them.

Moazeni, S., Powell, W.B., Hajimiragha, A.H., 2014 (in press). Mean-Conditional Value-at-Risk Optimal Energy Storage Operation in the Presence of Transaction Costs. IEEE Transactions on Power Systems

Energy Stored in the Wire?

by Emil Morhardt

How about storing the excess energy not chemically in batteries, not as potential energy in pumped hydroelectric reservoirs, not as compressed air, or heat, but just as excess charges? That's what capacitors do, and supercapacitors are widely used for storing not-too-much energy for immediate release when needed. What if the electrical cables themselves could be turned into supercapacitors. Yu and Thomas (2014) figured out how to do just that by using nanowires spaced around the central charge carrying cable to accumulate the excess charge. If wires like this become economic, they could be used to connect photovoltaic panels and wind turbines to the grid, automatically leveling out the delivery of power when the source becomes intermittent.

Yury Gogotsi writes about it in the 29 May 2014 issue of Nature

Fast Discharge Batteries: Electric Eels
by Emil Morhardt

Though this has more to do with generating electricity and using it wisely than storing it *per se*, one of the oldest electric batteries is the 600 Volt biological battery of the electric eel *Electrophorus electricus*. These animals search about for hidden prey by emitting a couple of high voltage pulses from time to time to see if anything jumps. If it does, they sidle on over and cut loose a volley of high frequency (~400 Hz) pulses that the muscles of the prey apparently interpret as coming from their own nervous system. The result is many muscles contracting simultaneously, paralyzing the prey into a state of whole-body muscle contraction known as tetanus (similar to the eponymous disease) and the eel sucks them in. This all happens pretty fast, on the order of milliseconds. If the eel fails to suck them in they often just swim away.

In a series of elegant experiments, Kenneth Catania (2014) at Vanderbilt University explored this in some detail by isolating immobilized fish behind an electrically-transparent barrier—keeping the eel at bay—while he measured the eel's volleys and the prey's responses. Given that everyone knows electric eels exist, and that electrophysiologists can't help being interested in them, it is somewhat surprising that all this wasn't known before. Fisheries biologists know all about the effect of shocking fish because they use electrofishing gear to temporarily stun fish so they can be captured and counted, and I had always assumed it was doing something similar to a hunting electric eel. Dr. Catania and the editors of science interpret this phenomenon in a more nuanced way, characterizing the eels' behavior as "...remotely control[ling the] muscles of their prey." Somehow, viewed in that light, the process seems far more nefarious than simply stunning the prey.

Catania, K., 2014. The shocking predatory strike of the electric eel. Science 346, 1231-1234.

Battery Imaging Sheds Light on Future Renewable Energy

by Niti Nagar

Researchers at the University of Wisconsin-Madison have developed an innovative X-ray imaging technique to visualize and study the electrochemical reactions in lithium-ion rechargeable batteries. Unlike usual batteries, these contain iron fluoride. "Iron fluoride has the potential to triple the amount of energy a conventional lithium-ion battery can store," says Song Jin, a UW-Madison professor of chemistry and Wisconsin Energy Institute affiliate. "However, we have yet to tap its true potential."

The team experimented with a transmission X-ray microscope at the National Synchrotron Light Source. They produced a chemical phase map that tracks the electrochemical discharge of iron fluoride microwires. In the past, it was difficult to understand what happened to the iron fluoride because other battery components hindered a precise image. By accounting for the background signals that would otherwise confuse the image, the team was able to accurately visualize and measure, at the nanoscale, the chemical changes iron fluoride undergoes to store and discharge energy.

Using iron fluoride in rechargeable lithium ion batteries has presented scientists with two challenges. The first is that it doesn't recharge very efficiently. However, by examining iron fluoride transformation in batteries at the nanoscale, Jin and his team's new X-ray imaging method pinpoints each individual reaction to understand why capacity decay may be occurring. The second challenge is that iron fluoride battery materials are not energy-efficient, meaning they do not discharge as much energy as they take in. Jin and his team's study shed some insights into this problem and plan to further investigate this challenge in future experiments.

Jin sees a big and broad range of applications of this technology. In addition to using portable electronic devices for longer before charging, Jin says, "we could advance large-scale renewable energy storage technologies for electric cars and micro grids." He also believes that the novel X-ray imaging technique will facilitate the studies of other technologically important solid-state transformations and help to improve processes such as preparation of inorganic ceramics and thin-film solar cells.

Linsen Li, Yu-chen Karen Chen-Wiegart, Jiajun Wang, Peng Gao, Qi Ding, Young-Sang Yu, Feng Wang, Jordi Cabana, Jun Wang, Song Jin. Visualization of

electrochemically driven solid-state phase transformations using operando hard
X-ray spectro-imaging. Nature Communications, 2015
DOI:10.1038/ncomms7883
University of Wisconsin-Madison. "Better battery imaging paves way for renewable
energy future." ScienceDaily. ScienceDaily. 20 April 2015.
<www.sciencedaily.com/releases/2015/04/150420182411.htm>.

California's Lack of Storage for Clean Energy
by Shannon O'Neill

In April of 2014, a surplus of renewable energy prompted officials in California to ask wind and solar plants to cease their operations for 90 minutes. During this time, the renewable energy firms reduced their output by 1,142 megawatts due to lack of storage for the energy. This resulted in enough clean energy to fuel thousands of homes being wasted. Meanwhile, dirty energy from nonrenewable sources continued to supply homes throughout this time due to their ability to stay stable on the energy system.

This issues stems from the fact that the peak demand for electricity in homes seldom coincides during times of the day with the most wind or sunshine. The lack of storage for clean energy has therefore resulted in loss of renewable energy and has hindered the California's efforts to rely more on renewable energy.

In efforts to deal with this issue, California is requiring three of the largest utility firms to invest in hundreds of megawatts of storage over the next several years. However, if the state hopes to reach Governor Brown's proposal of half of the state's energy coming from renewable sources by 2030, this storage will not be enough.

Companies are researching and developing supersized batteries in order to store energy that can be used to store and then provide energy when the demand is high. Additionally, officials are working with grid operators throughout the West in hopes of spreading clean energy to areas in need in order to reduce the usage of dirtier fuels, while simultaneously ensuring that clean energy is not going to waste.

Specifically, a project in Vacaville, California has two, 2-story metal boxes that contain battery cells with the capability of storing 2 MWh of electricity. This is enough energy to provide power to 1,400 homes per day. However, such projects like this have been problematic as they are very expensive—this one costing $3.3 million—and rely on taxpayer money in order to get started.

Such research and development in the storage of clean energy is expanding rapidly in the right direction. The manager of SunEdison believes that storage systems could soon be capable of storing hundreds of megawatts of clean electricity, fueling the transition away from nonrenewable energy sources.

LA Times: "California's push for clean energy has a problem: no place to store it" (http://www.latimes.com/local/politics/la-me-pol-green-energy-20150324-story.html#page=1)

California Needs to Invest in Renewable Energy Storage Technology
by Melanie Paty

Governor Jerry Brown is committed to promoting renewable energy development in California. He recently mandated that the state's major energy utilities install 1,300 MW of storage by 2024, however, grid operators say that it won't be enough to handle the energy input generated if the state meets its target of 50% renewable energy by 2030. Just last April, wind and solar operators were forced to cut their output when supply exceeded demand, while fossil fuel energy was allowed to stay in the system to keep it stable. During this 90-minute period, the amount of clean energy lost could have powered hundreds of thousands of homes. Government officials are hoping that the storage mandate will encourage companies to invest in more storage technologies. As of now, the main storage technology in development is the battery. A Pacific Gas and Electric facility is experimenting with metal boxes filled with battery cells that store 2MW of electricity. While this is an insignificant storage quantity, if this technology is successful, developers are hoping to expand it across the state. Berkley-based SunEdison is buying Solar Grid Storage, a start up, which develops batteries that absorb sun energy when it is sunny and there's limited demand. LightSail is developing a system that uses green energy to pump compressed air into storage tanks; when demand is high, the air is released through a turbine, generating electricity. However, these products are still expensive and rely heavily on tax dollars and private funding. EnerVault's recent battery prototype project cost $10 million, half of which came from the US Department of Energy. However, isolated batteries are not the only solution to storage. Governor Brown sees potential in electrical cars charging when demand is low and supply is high. Clean power could also be exported to energy grids of neighboring states—a method Europe has found successful.

Megerian, Chris. LA Times, "California's push for clean energy has a problem: no place to store it." March 23, 2015. [http://www.latimes.com/local/politics/la-me-pol-green-energy-20150324-story.html#page=1]

Lithium-Air Batteries: The Next Battery Revolution?

by Chad Redman

Energy storage technology is a fascinating field of research, boasting numerous potential replacements for the current cutting edge of battery technology – lithium-ion chemistry. One promising and widely researched alternative to lithium-ion batteries is the lithium-air battery. Lithium-air battery technology became widely known to researchers in the field in 2009, and has been the subject of over 300 research papers since 2011. The primary benefit of lithium-air technology is increased energy density; in fact, lithium-air battery technology has the potential to bring electric vehicle range up to competitive levels with internal combustion engines (Girishkumar *et al.* 2010). Lithium-air batteries with non-aqueous electrolytes (oxygen gas) can theoretically produce up to 3500 watt hours of energy per kilogram, or 1700 Wh/kg in practical application (Kwabi *et al.* 2014).

However, as it stands, lithium-air batteries are far from commercial development. There are large challenges which render this technology unusable for the time being. Specifically, two primary challenges include the reactivity of lithium with water, and the recharging of the pack by reversing the lithium-oxygen reaction. Current materials are insufficiently robust and reliable for preventing water from penetrating the lithium housing in lithium-air batteries, and merely a tiny amount of water finding its way to this store of lithium would result in critical failure. Resolving this challenge requires the development of an electrolyte that will permit only the selected reagents access to the lithium electrode. A second major challenge to lithium-air researchers is the rechargability of these batteries. Little progress has been made on this front, and researches continue to look for configurations of lithium-air batteries that will allow for the reversal of the lithium and oxygen reaction.

Ingram, Antony. "How Hard Is Lithium-Air Battery Research? Pretty Tough, Actually." Green Car Reports. May 8, 2014.

Kwabi, D.G., Ortiz-Vitoriano, N., Freuneberger, S.A., Chen, Y., Imanishi, N., Bruce, P.G., and Shao-Horn, Y. (2014). Materials Challenges in Rechargeable Lithium-Air Batteries. Materials Research Society, 39, 443-452.

Girishkumar, G., McCloskey, B., Luntz, A. C., Swanson, S., and Wilcke, W. (2010). Lithium-Air Battery: Promise and Challenges. American Chemical Society.

Liquid Air Energy Storage
by Chad Redman

Asymmetrical energy production and consumption over the course of the day creates challenges all around the globe, which is why effective and efficient energy storage technologies are the subject of widespread research and development. Liquid Air Energy Storage (LAES) is one fascinating method for storing excess, cheap off-peak energy, and taking advantage of it when energy production falls and demand rises in the evening. The Energy Storage Association describes the systems behind LAES, including ways in which waste from unrelated processes can be turned into valuable energy.

LAES utilizes super-cooled air or nitrogen, transformed from a gas into a liquid, to store energy in insulated tanks. The first stage of the LAES system is dubbed the "charging" stage, and involves the intake of ambient air and the cooling of that air till it assumes a liquid form. The second LAES stage is energy storage, usually holding the liquid air in tanks similar to those used for liquid nitrogen, oxygen, and LNG. Finally, the third stage of LAES is the "power recover" stage, in which the liquid air is evaporated and heated to ambient temperature. This process raises the pressure of the air, which can then be used to run a generator.

LAES has several characteristics that make it attractive in the long run as a money and energy saving device. First, the hardware components required to construct a functional LAES system are highly prolific, and boast reliably long lifespans. Furthermore, there are a few processes that can be used to boost the efficiency of LAES systems. First, very cold air produced from the third stage of LAES can be stored and reused to liquefy air on the next cycle. Cold air from other industrial processes can be utilized similarly. Similarly, this situation is true in reverse; hot air produced in the first stage of LAES can increase efficiency of the energy recovery stage. Waste heat from external industrial processes can be used in the same way.

"Liquid Air Energy Storage (LAES)." EnergyStorage.org. N.p., n.d. Web. 02 Feb. 2015.

Redox Flow Batteries: For Grid Level Storage

by Chad Redman

Current energy storage technologies are often overlooked in favor of the next promising development that will be commercialized sometime in the future, but economical large scale energy storage is already possible with current equipment. Redox flow batteries (RFBs) are a type of large battery that utilizes reduction and oxidation reactions to charge and discharge liquid electrolyte solutions. The advantage of RFBs over other battery types is realized in scale; RFBs can easily expand and store more energy by using larger storage tanks for the electrolyte solutions. However, the power that can be produced by an RFB is determined by the architecture of cells within the RFB. Unlike a standard Li-ion or lead-acid battery, only a small percentage of the energy within an RFB is accessible as power at any given moment.

RFB development has produced three separate flow battery technologies in recent years. The first of these is the Iron-Chromium (ICB) flow battery. This battery type employs iron and chromium as electrolyte fluids, and has the advantage of being reliable relative to other RFBs. However, ICBs come up short on the power density front, making them less promising in the long run. Interestingly, the Electric Power Research Institute supported ICB technology in 2010, showcasing its safe, reliable, and cost-effective nature.

Two other RFB types are under development, each boasting much higher energy densities than ICBs, but each being less reliable, less safe, and more expensive to operate. Both vanadium redox flow batteries and zinc-bromine flow batteries are capable of being charged to such a power-dense state that the cells and membranes within the batteries deteriorate and fail. As these technologies improve, they will replace ICBs as a more effective form of RFB.

"Redox Flow Batteries." EnergyStorage.org. N.p., n.d. Web. 09 Feb. 2015.

Flywheel Kinetic Energy Storage: Energy in Motion

by Chad Redman

Rapidly spinning masses known as flywheels are used for energy storage in a wide variety of applications, including transportation, sport, and grid level electricity. Focusing on grid solutions, flywheel energy storage systems (FESS) comprise massive rotors magnetically suspended in a stator, which acts as a motor when the flywheel needs to be spun up and a generator when the kinetic energy of the flywheel needs to be converted into electricity. Through the use of magnetic bearings and a vacuum chamber for the flywheel housing, FESS are highly efficient for short-term energy storage.

FESS have a set a characteristics that make them very effective for specific applications, but impractical for operations at the largest scale. These flywheels produce very high power electricity, capable of doing work quickly. However, they do not store tremendous volumes of energy, so they are best suited to feed applications that require high power but low energy. Moreover, FESS can be run through well over 100,000 cycles of full charge-discharge, giving them better longevity than most energy storage technologies and making them a fitting solution for high frequency processes. This advantage is further supported by the quick ramp rates of FESS; flywheels can achieve full energy storage in a matter of seconds if excess power becomes available. Unfortunately, while this high-tech solution is an environmentally sound option for energy storage, it is also highly expensive relative to competing substitutes.

"Flywheels." EnergyStorage.org. N.p., n.d. Web. 16 Feb. 2015.

Pumped Hydroelectric Storage: Putting Gravity to Work
by Chad Redman

Damming natural flowing rivers is an ancient and effective method for generating renewable energy. However, sufficient rivers are a scarce resource and modern dams produce an array of undesirable environmental effects. In response to the drawbacks of traditional dams, the main commercial technique for storing potential energy in water is pumped hydroelectric storage (PHS). Traditionally, these facilities use a massive pump and two reservoirs, one elevated above the other. During off-peak hours, excess energy produced from sources such as wind farms and nuclear power plants is used to power a pump which moves water into the elevated reservoir. When energy demand rises, the water is released back into the lower reservoir, spinning the pump which effectively becomes a generator.

Two major forms of pumped hydroelectric storage exist. Open systems, which make up the vast majority of current PHS systems, use a body of free flowing water for one or both reservoirs. They do not require an initial filling of the system, the construction of artificial reservoirs, or perpetual compensation for evaporating water. They do, however, impact marine ecology. Closed loop systems, on the other hand, involve two reservoirs isolated from flowing water and allow for the use of waste water, such as grey water. These installations are inherently isolated from aquatic life.

Another advancement toward more efficient and effective PHS is the use of variable speed pumps. Historically, the transfer pump between reservoirs operates only at two speeds – on or off. This has meant poor matching of pump energy use with actual excess energy available, as well as lack of frequency control when the motor is used to generate electricity. Variable speed pumps solve these problems, allowing the pump to use precisely the excess energy available, as well as endowing the PHS facility with a tool for regulating frequency when in generation mode. This creates a safer and more efficient technology. However, variable speed pumps are new to the game and come with a couple of drawbacks. First, variable speed pumps require extra machinery expense, specifically in more advanced rotors and additional frequency converters. Second, variable speed pumps typically exhibit high parasitic loads compared to fixed speed pump systems. Given the overwhelming benefits of variable speed motors for PHS systems, it is likely that future installations will find solutions to the shortcomings of this technology.

"Pumped Hydroelectric Storage." EnergyStorage.org. N.p., n.d. Web. 23 Feb. 2015.
"Variable Speed Pumped Hydroelectric Storage." EnergyStorage.org. N.p., n.d. Web. 23 Feb. 2015.

IMPROVING THE GRID

There are many ways in which the world's electrical grids—the network of power lines and associated hardware—can be improved, and many of these are collectively referred to as "smart" grids. "Smart" electrical meters are a start and these have been widely installed in developed countries, communicating instantaneous power usage to the grid operators and potentially effecting demand-side energy regulation—in effect, depriving users of as much energy as they would like in ways which are largely unnoticeable to them. Another important feature will be adding storage to the grid so that it can accommodate much more renewable energy. Not much has transpired along these lines just yet, but an improved grid is high on energy suppliers' wish lists as a way to meet demand in the absence of new generation facilities. In parts of developing countries (and even in parts of the US, there is essentially no grid at all, so an alternative is micro-grids, serving small isolated communities. Much more will be written about these issues in later editions of this book, as they mature, but for now we have a few interesting examples.

New York Plan to Open a Smart Grid Laboratory

by Mariah Valerie Barber

As part of New York's Reforming the Energy Vision (REV) plan, this past March, Governor Andrew Cuomo announced the state of New York's plan to build a smart grid research laboratory, called the Advanced Grid Innovation Laboratory for Energy. The new laboratory is being built in response to the damage Hurricane Sandy did to New York's power system. At one point during Hurricane Sandy over 8 million residents went powerless. In addition to the outages, the hurricane caused billions of dollars worth of infrastructural damages to the city. Hurricane Sandy was a harsh representation of global warming and rising sea levels.

In order to prevent such damage from being done to New York and to become better prepared for the changing climate, the state of New York will build a new advanced power grid that incorporates renewable energy into it. The new power grid will aim to end the dependency of small or individual power plants on centralized power systems. Instead the new grid will enable independent power plants to supply smaller local systems, allowing them to be self sufficient. Such independent power generators are known as microgrids. Examples of local systems include certain residential areas, businesses, and hospitals. By utilizing microgrids the state could prevent a far-reaching and large-scale power outage like the one that resulted from Hurricane Sandy.

These more self-sufficient microgrids will be powered by small wind turbines, solar panels, other renewable energy sources, or even diesel, natural gas, or batteries.

In addition to making local systems less dependent upon centralized power generators, the research lab will utilize smart grid technology, which is a computerized grid that allows the efficiency of electricity use to be maximized. It does so by enabling the communication between "utility and costumers and the utility and power plants, enabling power to be used more efficiently, and allows wind and solar power plants to be integrated into the system" (Climate Central). This communication can prevent excess electricity use or production. Power systems, utilizing smart grid technology can rely on both centralized and renewable energy, and in the case of a blackout on the end of the centralized energy, the entity dependent on the electricity could switch to the other power source reliant on renewable energy. As New York works to develop and put to use such a grid, other parts of the United States are taking notice. If this grid system is deemed successful it could act as a model for many other parts of the nation (Climate Change).

Climate Central (http://www.climatecentral.org/news/new-york-smart-grid-r

Microgrid Micromanagement
by Briton Lee

One of the issues with the integration of alternative energy, such as solar and wind power, into the electricity grid is their volatile load swings. Automation and microgrids seek to address this issue of fluctuating energy and make renewables more amenable to integration.

Solar and wind power are unpredictable, and fluctuations occur simply when a cloud passes over a solar grid. Another problem is that solar energy is generally produced during the day and not during the night, whereas human electricity use peaks in the evening. Generally, humans have to manually monitor and balance energy production and consumption in order to manage the electrical loads. The entire grid is tightly monitored, and the formulas used to keep the grid in check are thrown off when renewables are included. Renewables are unpredictable because it's unclear when the energy will come in, since energy is not stored but rather threaded directly into the grid.

Microgrids are smaller distributed grids, and act as a reserve that manages the fluctuation of alternative energy sources. They use energy sources and storage that can handle smaller fluctuations on their own. The microgrids are implemented on both the supply and demand side; they are able to communicate with each other to determine how much energy is needed and compares it to how much energy is produced. In a microgrid controller system, different energy loads are connected to a control node, which consists of a communication node, power measuring device, and a switch that can isolate the loads leading to a more controlled and automated process of energy management.

These microgrids present huge savings in management costs and makes the adoption of renewables more palatable. Microgrids aren't exactly solving fluctuation issues, but they make it so that grid engineers do not have to concern themselves with smaller issues, and can instead focus on fluctuations that exceed 10 MW. Automation in the renewable industry gives alternative energy a stronger footing to tackle the hegemony of fossil fuels.

Spiegel, Rob. 2015. "Automation Gives Alternative Energy a Boost".
(http://www.designnews.com/author.asp?section_id=1386&doc_id=276789).
Zaidi, A., et. al. 2008. Microgrid automation - a self-configuring approach. Multitopic Conference, 2008. INMIC 2008. IEEE International.

Consequences of the Electric Grid's Inefficiency
by Shannon O'Neill

Solar Panels are usually installed facing south in order to capture the most energy from the sun. This placement is mandated under the rules that govern the electric grid, as solar panel owners get paid for the amount of energy that their panels generate. This has resulted in panels capturing energy when it is least needed, providing insight to how such rules not only create more costs but also reduce effectiveness.

The greatest demand for energy comes during late afternoon, when temperatures are higher and many people are inside running their air-conditioning units. During this time, the placement of the panels causes the sun to hit at an indirect angle, reducing the amount of energy harnessed. This has led some experts to suggest that the panels should be placed facing west—toward the setting sun—in order to capture the sunlight energy when it is needed most. However, under current rules, this would also result in a decline of total production that will reduce profits for panel owners.

Resolving such issues is necessary in order to transition away from natural gas. Natural gas has been able to keep its place on the market due to the fact that it can supply energy on demand. Renewable energy sources often flood the market with energy when it is not needed, pushing prices down, sometimes even causing grid operators to be charged for adding more energy to the system than needed.

This becomes specifically troubling for nuclear plants, as they cannot quickly alter their output, resulting in such firms being fined for excess production. This has reduced the income of some nuclear plants, as renewable energy and cheap natural gas have lowered energy prices and are used preferentially by many utilities.

Though the nuclear firms are being charged for the excess energy, current policies in some jurisdictions allow wind farms to continue to earn tax credit for each hour of energy they make whether it is needed or not. Therefore, the firms that are responsible for flooding the market with energy continue to make profits at the expense of the more constant, reliable energy sources. Joseph Dominguez, the senior vice president of Exelon, a nuclear utility, states that this has left us in a place where "we devote so many of our economic resources and policies to the type of energy that produce power but not power on demand, [leaving us] in a place where we start losing the megawatt we can control" (New York Times). This inefficiency and loss on control is hindering the process of the complete transition away from fossil fuels.

New York Times: "How Grid Efficiency Went South"
    (http://www.nytimes.com/2014/10/08/business/energy-environment/how-grid-efficiency-went-south-.html?ref=businessspecial2)

The Future of Electricity: Batteries and the Grid
by Ali Siddiqui

Many people believe that wind and solar energy is the future, and that fossil fuels are the past. California has set a goal of deriving 33% of its electricity from renewable energy by 2020, and while many people applaud this initiative because of the great reduction of greenhouse gasses, some share a concern that solar and wind energy each have highly fickle natures and that uncertainty can cause strain on a power grid. Batteries provide a solution to this variable nature. Owner's of batteries can charge their batteries during times of high available energy in the grid and store that energy until times of low energy on the grid. When the supply of energy is low, then the owners of those batteries can sell back their stored energy because demand for that energy will be high. The supply of energy will be low during times the sun doesn't shine or the wind doesn't blow, and this would allow that variable cost associated with renewable energy like solar and wind to be reduced.

Ravi Manghani, an analyst with GTM Media, has stated, "we're going to see on an average of 100 to 250% growth" and "most of that will be in batteries." However, even with these predictions, people still do not posses large battery systems, and the utilities and companies that do are using them for different purposes. Battery systems are rows of cells strung together capable of storing large amounts of power and releasing it over an hour or longer.

Greg Wolf, president of Duke Energy Renewables, owns a large battery system in Notrees, Texas. He has, however, stated that the energy he has stored has had little interest or demand even from neighboring wind turbine farms. Storage of energy like this may be the future, but this future is still a while away.

Fountain, Henry. Batteries and Renewable Energy Set to Grow Together. April 20th 2015. http://www.nytimes.com/2015/04/21/science/batteries-and-renewable-energy-set-to-grow-together.html?_r=1

Europe's Power Grid to be Increasingly Interconnected: Norway Plans Deep-Sea Cables to Germany, England
by Trevor Smith

Norway is in the process of finalizing plans to build massive submarine power cables to link its power grid to England's and Germany's grids. The move is being praised as a win for clean energy, as the cable will allow for exporting excess hydroelectric energy from Norway to England and Germany. The cable to Germany is set to be completed by 2018, while the cable to England will be finished by 2020 (Reuters 2015).

The nearly four hundred mile cable between Germany and Norway, christened NordLink, will be capable of supplying up to 3% of Germany's energy. How much energy Germany will actually draw from Norway is dependent on how much energy Germany is producing itself—the cable is designed to account for fluctuations in Germany's solar and wind power production and draw power from Norway only when necessary (Rueter 2015). The cable can also be used to transfer energy in the opposite direction, sending any excess energy Germany produces to Norway.

This power exchange system has the potential to become even more powerful with the addition of cables connecting other countries with Norway. A 435-mile cable connecting Norway and England is expected to be completed by 2020. The deal is very similar to that between Germany and Norway; England plans to import excess hydropower energy from Norway and to offload their own excess wind and solar energy through the cable as well (Reuters 2015). After the completion of the cable, England will then be connected to Germany in addition to Norway. The end result will be a much more stable renewable energy grid. Because dips or spikes in power production in any of these three countries can be offset by pulling power from any of the others (even if only indirectly through Norway, in the case of England and Germany), produced power can be more consistently and efficiently used. The cables do come at a cost, however: The cable to England will cost 2 billion euros, while the cable to Germany will cost approximately 1.5 billion euros.

Gero Rueter. "Norway to become Germany's energy reservoir" DW, February 16, 2015. http://www.dw.de/norway-to-become-germanys-energy-reservoir/a-18256802
"Norway's Statnett sees final decision on 2 bln euro UK cable in Q1" Reuters, Jan 5, 2015. http://uk.reuters.com/article/2015/01/05/norway-uk-cable-idUKL6N0UK1Y720150105

Crossing Borders Through Electricity

by Abigail Wang

What better way to connect countries than through renewable energy? As countries look towards being more environmentally friendly and conscious, they're trading energy to satisfy the needs of their own people and their neighbors. The United Kingdom and Norway plan to build the world's longest undersea interconnector to provide low-carbon energy for nearly 750,000 British homes.

An interconnector is a connection between the electricity transmission systems of different countries, in this case through underwater cables. The proposed interconnector is a two-way, 1,400-megawatt electricity cable that will run from Blyth in Northumberland on the UK side to Kvilldal in Rogaland on the Norwegian side. When winds are strong in the UK and the wind power production is high, Norway will be able to import power from Britain at a lower price than could otherwise be found in the Norwegian market. This will help Norway conserve water in their hydropower reservoirs. When the situation is reversed and wind is slow with a greater need for power in the UK, Britain can import Norwegian hydropower. The interconnector will save UK households up to three and a half billion pounds over 25 years. It will also increase the security and predictability of power supply for both countries.

National Grid, an international electricity and gas company based in the UK, and Statnett, a Norwegian transmission system operator, signed an ownership agreement in which the two companies will split the cost of the 450-mile interconnector. The project will cost about two billion euros and completion is slated for 2021.

This is the first electricity link between the UK and Norway, but these countries have been energy allies for a long time. In the past, their partnership revolved around fossil fuel resources in the North Sea, but the two now want to share renewable power. One of their most well known projects, The Langeled Pipeline, was completed in 2006. It runs from the Nyhamma terminal in Norway to Easington in Yorkshire, England and transports Norwegian natural gas to the UK.

The UK already has electricity interconnectors with France, Ireland, and the Netherlands, and adding Norway to the list will help the country meet its expected energy demand in the upcoming years. Norway is also increasing its energy partners; last year, Statnett laid a similar interconnector between it and Denmark. Sharing renewable energy could help countries meet the challenge of finding greener, affordable energy.

Farrell, Sean. "UK and Norway to Build World's Longest Undersea Energy Interconnector." The Guardian: 26 March 2015. http://www.theguardian.com/business/2015/mar/26/uk-and-norway-to-build-worlds-longest-undersea-energy-interconnector

"Norway and National Grid Announce World's Longest International Electricity Connector to UK." Scottish Energy News: March 2015. http://www.scottishenergynews.com/norway-and-national-grid-announce-worlds-longest-international-electricity-connector-to-uk/

NOVEL ENERGY APPLICATIONS

Some of the most interesting aspects of energy research and development are new applications that often seem to come out of the blue. This section describes many of the most exciting ones.

ORNL Creates Low Cost Energy Sensors
by Mariah Valerie Barber

Oak Ridge National Library, the largest US Department of Energy science research laboratory has created new low-cost wireless sensor technology that can be used to monitor the energy consumed by commercial buildings (Ornl.gov). Currently, buildings consume 40% of all energy being consumed in the United States. Most commercial buildings poorly monitor and control their energy consumption. For example, systems in commercial buildings such as heating, ventilation, air conditioning, and electricity often are under controlled and unmonitored. These new sensors have the potential to reduce the energy consumption of buildings by 20–30% (Physics.org).

The sensors use technology that prints circuits, sensors, antennae, and photovoltaic cells and batteries onto very thin and flexible plastic sheets with adhesive peel-and-stick on the back of it. In addition to being able to be printed and installed in buildings very easily, the sensors are extremely low-cost. The sensors require very little power and are entirely wireless. They are to be stuck with the adhesive glue on the back of them to various walls throughout the building. They monitor for outside air, room temperature, humidity, light level, occupancy, and pollutant. They collect data on each factor that they are monitoring and then send that data a main receiver, which receives data from all the other sensors placed in the building. Since they are wireless they are extremely easy to install and place throughout a building.

These ORNL sensors are extremely cheap, ranging from $1-10 per sensor. Currently, the wireless sensors that are available commercially cost around $150 to $300.

As of now the ORNL's sensors are not available for purchase or commercial use, but the ORNL and the US Department of Energy are in the process of negotiating with developers and international electronic manufactures in order to create an agreement so that these sensors can become widely available commercially (Physics.gov).

Oak Ridge National Laboratory. (http://www.ornl.gov/about-ornl)
Physics.org (http://phys.org/news/2015-03-sensors-yield-energy-efficiency.html)

Blue Energy Pilot Plant Opens in the Netherlands
by Mariah Valerie Barber

On November 26th, 2014, the world's first "blue energy" pilot power plant opened up in the Netherlands. The plant is being overseen by the company REDstack BV, founded in 2005, which utilizes Reverse Electro Dialysis (RED) to create energy (Energising Deltas). With the newly opened blue energy pilot plant, REDstack BV and the Dutch government hope to learn enough so that the technology can be improved and so that blue energy plants can open commercially by the 2020. The blue energy plant was constructed at the Afsluitdijk causeway, in the Netherlands, that stretches for nearly 20 miles. The Afsluitdijk was constructed in 1930, turning half of the North Sea into a freshwater lake. On one side of the Afsluitdijk causeway is the Ijsselmeer, the man- made freshwater lake and on the other side is the Wadden Sea, which is part of the North Sea. Constantly at Afsluitdijik, freshwater of the Ijesselmeer and salt water of the Wadden Sea are coming in contact with each other, which is why it makes the perfect site for the world's first blue energy plant; blue energy runs off this exact occurrence. Blue energy comes from Reverse Electro Dialysis (RED) energy that is retrievable from the differences in the salt concentration in salt water and freshwater. These differences in concentrations can be used to produce electricity (University of Twente). The power plant utilizes membranes made by Fujifilm, or two part filters in order to conduct RED. One of the filters allows the positively charged sodium ions of the salt waters to enter and other filter lets in negatively charged chlorine ions. With both the positive and negative charges from the salt and fresh water a natural battery is created, and subsequently, electricity (Fujifilm.com).

Professor Kitty Nijemeijer of MESA + research institute of the University of Twente, who has been researching blue energy for several years, stated,

"This is because there are many more charged particles—ions—in salt water than in fresh water. Separating salt water from fresh water, using a membrane that only allows positively or negatively charged particles to permeate, results in a difference in voltage which can be converted into electricity. The principle has been known for some time, but its efficiency had always been far too low to make large-scale application interesting" (University of Twente).

Currently, at the new test plant, 220,000 liters of salt water and 220,000 liters of fresh water can be processed every hour. Each square meter of membrane is capable of producing 1 watt and the researchers hope to be able to increase that amount to 3 watts with the knowledge that comes from running the test plant. Eventually the plant should be able to produce 50 MW (Physics.org).

University of Twente.
   (http://www.utwente.nl/en/newsevents/!/2014/11/349878/first-blue-energy-power-plant-opened-ut-made-large-contribution)
Physics.org. (http://phys.org/news/2014-11-dutch-harness-energy-salt.html)
Fujifilm.com (http://www.fujifilmmembranes.com/blue-energy-technology)
Energising Deltas.com (http://www.energisingdeltas.com/partners/redstack/)

New Printable Circuits Change the Future of Wearable Tech
by Alex Elder

A new technology developed at Purdue University could change the way that we think about wearable technology. Although innovations like Google Glass and the Apple Watch have boosted the public's interest in wearable technology, these items are essentially just smaller versions of existing machinery. New research at Purdue University revealed that inkjet-printing can be used to mass-produce electronic circuits made of liquid-metal alloys on any surface. This advancement could change the way wearable technology is used all together. Stretchable garments with integrated electronic technology could offer the same types of advantages as current wearable tech but without any movement restriction and with more versatility. Additionally, pliable robots that can change shape and fit through small spaces could become a reality.

The technology operates by using traditional inkjet printing processes to print flexible and stretchable conductors onto almost any surface, including elastic materials and fabrics. The conductors, which are made from a metal alloy, are able to stretch and deform without breaking or losing their functionality. The printable ink is made by dispersing the liquid metal in a non-metallic solvent like ethanol using ultrasound. This process breaks up the bulk liquid metal into a nanoparticle form which is compatible with inkjet printing. After printing, the nanoparticles must be rejoined by applying light pressure to make the material conductive. This approach makes it possible to select which portions of the circuit to activate, depending on the design.

However, before this technology can be incorporated at the commercial level, new manufacturing techniques must be developed in order to mass-produce these liquid-metal circuits. Further research is also necessary to explore how the interaction between the ink and the surface being printed on might affect the production of specific types of devices.

Moon, M. (2015). Liquid mental printing puts flexible circuits on 'anything.' Engadget. http://www.engadget.com/2015/04/08/purdue-inket-printer-liquid-metal/

Venere, E. (2015). Inkjet-printed liquid metal could bring wearable tech, soft robotics. Purdue University. http://www.purdue.edu/newsroom/releases/2015/Q2/inkjet-printed-liquid-metal-could-bring-wearable-tech,-soft-robotics.html

Extending the Range of Wireless Charging
by Briton Lee

Wireless charging has been quite prominent in recent years, and the implications of the technology are obvious, bypassing the need for any physical connection to power any device with energy. However, the technology is still limited and in its early stages, and the implementation of wireless energy is highly restricted by distance, as it was when it was initially introduced in 2009 with the Palm Pre (Miller 2009). Currently, this technology is seen mostly in wireless charging mats that are only slightly more convenient than plugging in a device. However, during the 2015 Consumer Electronics Show (CES), the company Energous debuted their work on long range wireless charging, WattUp. The technology attempts to achieve the goal of never having to worry about charging devices again. While this extent of wireless charging has been introduced before, there were problems with being able to scale up the efficiency. However, Energous has come a long way since it was established in 2012 as WattUp is able to achieve over 70% efficiency (charging mats are approximately 90% efficient)

(Souppouris 2015). The way the charging works is similar to how Wi-Fi works—the WattUp router sends out a radio wave that can be converted by proprietary receivers back into usable electrical energy. These receivers are relatively small, and can be integrated into just about anything by electronic manufacturers. For example, at CES they were able to demonstrate the charging capabilities of the router using a phone with an extended case containing the receiver. It might seem that outputting such a level of energy constantly would have adverse effects, but since it is the same technology as Wi-Fi, it can be extrapolated that there will be no adverse effects related to this technology. There may be an additional worry that WattUp wastes energy if energy is constantly being output, but the receivers and router communicate with each other via Bluetooth to determine whether it needs to be charged. There is a lot of promise in this developing technology, but it may have some unforeseen consequences. It is possible that it will contribute to needless charging of devices, and promote the idea that energy is limitless as we become more and more removed from where our energy comes from.

Souppouris, Aaron. 2015. Engadget. "This router can power your devices wirelessly from 15 feet away". (http://www.engadget.com/2015/01/05/energous-wattup-wireless-charging-demo/).

Miller, Paul. 2009. Engadget. "Palm Pre's wireless charger, the Touchstone". (http://www.engadget.com/2009/01/08/palm-pres-wireless-charger/).

Energous: WattUp™ Overview. (http://www.energous.com/overview/).

Peanuts Packing a Punch
by Briton Lee

Though packing peanuts are fantastic for shipping applications, they're environmental nuisances that are difficult to recycle. At Purdue University, Professor Vilas Pol searched for something useful to do with the boxes of packing peanuts littering his lab, and ultimately found a way to convert them into eco-friendly anodes.

Creating these packing peanut anodes is relatively straightforward, and involves the process known as carbonization, which reduces a compound to its carbon makeup. Essentially, these packing peanuts are charred by heating at 500 °C to 900 °C in an inert gas, such as Argon, with or without a transition metal catalyst. This process is simple and benign, and has a positive environmental impact because it produces efficient electrodes while repurposing hard-to-recycle Styrofoam. In fact, the electrodes produced in this manner are more effective than conventional graphite electrodes; they have a measured specific capacity of 420 mAh/g (milliamp hours per gram), which is higher than the theoretical capacity of graphite (372 mAh/g).

Additionally, long-term use of the anodes is seen as particularly robust, with minimal capacity loss while maintaining integrity of the material's nanostructure after 300 full cycles (Venere 2015). One of the reasons that these packing peanuts-turned-anodes are so effective is that the carbonized polystyrene sheets are thin, compared to commercial graphene anodes, which can be up to 10 times thicker and lead to increased resistance and charging time. The method of creating these anodes makes them disordered and porous, which increases surface area and contact with the lithium electrolytes in the battery. Since the process of producing the anodes is so simple, it can be feasibly scaled up to industrial applications. The technique is versatile as well, since the styrofoam used is composed of polystyrene, the main component of plastics. The effectiveness of the material and its potential for application allows us to reuse styrofoam and other common materials effectively.

Venere, Emil. 2015. "New processing technology converts packing peanuts to battery components". Purdue University.

Torrice, Michael. 2015. "Pesky Packing Peanuts Baked and Crushed to Make Battery Electrodes". Scientific American.

Zimmer, Lori. 2015. "Researchers transform pesky packing peanuts into fast-charging batteries". Inhabitat.

Electricity From Low-Level Heat
by Emil Morhardt

Low-level heat—temperatures 100–200 °C above ambient, the temperature range of a kitchen oven more-or-less—are abundant in the exhausts of all sorts of industrial processes from drying biomass to operating internal combustion engines. They are also much more common in geothermal fields than the higher temperatures needed for traditional geothermal steam power generation, although low-level heat can be used to vaporize high-volatility organic compounds such as propane, which can then power a turbine much as steam would. For the most part, though, this heat is wasted, just released into the environment; but it needn't be. Researchers at the China University of Geosciences in Beijing and at Stanford University experimented with an array of commercially available thermoelectric power generators (TEGs) with the intent of producing a 500-Watt power source generated solely from low-level heat. The least expensive and evidently best TEGs for low level heat conversion are solid-state devices made of bismuth telluride, costing less that $3.50 per unit, and generating more that 4.5 W with a temperature gradient of only 140°C.

By combining 96 of these the engineers figured they could generate 500 W if they got the temperature up to 200 °C. This is a little more expensive than an equivalent output from photovoltaic panels in full sun, but has the distinct advantage that if the heat source is constant, as it would be in a geothermal situation or in an ongoing industrial process, the electrical output would be equally continuous. They may also be preferable to the propane-driven turbines now used in low-level geothermal electrical generation because they have no moving parts. It seems to me that they would be particularly valuable in northern climes where low ambient temperatures would allow generation from much lower-temperature geothermal or other sources, since it is the temperature gradient across the devices that determines the amount of electrical output.

Liu, C., Chen, P., Li, K., 2014. A 500 W low-temperature thermoelectric generator: Design and experimental study. International Journal of Hydrogen Energy. Abstract at: http://bit.ly/1sfVmtX

Bacteria Electrify Sewage, No Methane Needed
by Emil Morhardt

Bacteria are good at getting energy out of sewage; that's what wastewater treatment plants are mostly about...converting the organic carbon that we didn't extract from the food during it's passage through our guts into something that won't pollute the water bodies we dump the treated wastewater into. In the closed anaerobic digester tanks you can see at any wastewater treatment plant the microorganisms are busy converting it into methane. Sometimes this methane gets used onsite to generate power, or is further processed and piped off as "biogas" for some other use, maybe even to power city buses. More often than not it is just released into the atmosphere where, although it can no longer pollute any water, it is a powerful greenhouse gas. What if we could skip the methane production step and just generate electricity directly from the sewage by sticking electrodes in it? Sounds impossible, but there is new science that is making it happen, at least at laboratory scale. Xie *et al.* at Stanford University have constructed what they call a microbial battery that makes just as much electricity out of a given amount of wastewater as you can get from first using the microorganisms to produce methane, then burning it...without the intervening gas handling and power plant, not to mention the likely leaks of methane to the atmosphere in the process. The secret is a solid-state cathode which makes the system act like a rechargeable battery, with exoelectrogens—microorganisms that oxidize the electron-donating chemicals in the sewage and transfer the electrons to the anode. The electrons then pass through an external circuit as an electrical current, on their way to the cathode. Voila! Electricity that can be used for anything you like.

Xie, X., Ye, M., Hsu, P.-C., Liu, N., Criddle, C.S., Cui, Y., 2013. Microbial battery for efficient energy recovery. Proceedings of the National Academy of Sciences 110, 15925-15930. http://bit.ly/1siA7Ji

Microbial Electricity from Cyanide!
by Emil Morhardt

The paper by Xie *et al.* (discussed above) didn't say much about the electrogenic bacteria needed to make their microbial battery work. Researchers in Beijing and Singapore have published a paper focussed on such bacteria (*Klebsiella* sp. in this case), isolated out of a microbial fuel cell, that can do the job in wastewater heavily contaminated with cyanide, almost completely degrading the cyanide in the process. Even without the electricity generation this is interesting, because these bacteria do a better job of removing cyanide than the much more expensive chemical oxidation methods more commonly used by industry. Microbial fuel cells get electricity out of microbes differently than the "microbial battery" of Xie *et al.*; they consist of two wastewater-filled chambers seperated by a proton exchange membrane. The bacteria in the anode chamber strip protons (hydrogen ions) off the feedstock—a cyanide/glucose mixture in this experiment—and the protons migrate through the membrane to the cathode. The electrons flow as an electrical current from the anode to the cathode in a wire, where they can be used as electricity. Interestingly, the bacteria continued to generate electricity from cyanide alone when they ran out of glucose.

Wang, W., Feng, Y., Tang, X., Li, H., Du, Z., Yang, Z., Du, Y., 2014. Isolation and Characterization of an Electrochemically Active and Cyanide-degrading Bacterium Isolated from a Microbial Fuel Cell. RSC Advances, DOI: 10.1039/C1034RA04090B. Abstract at: http://rsc.li/XbkskA

Jumping-Droplet Electrostatic Energy Harvesting

by Emil Morhardt

Harvesting energy from atmospheric dew? Why would dew have any energy in it to harvest? Miljkovic *et al.* at MIT and Bell Labs, discovered it by accident while working with nanoengineered superhydrophobic surfaces (surfaces that *really* don't like water on them). When water condenses on them the droplets can merge and spontaneously jump off, and what's more, they are positively charged when they do. So, it's just a matter of collecting that charge and putting it in wires to get electric current that can do work. The authors capture the charged droplets on a superhydrophilic (really *likes* to have water on it) copper surface. The faster the droplets are going when they jump, the more power generation, and this can be achieved by getting them to jump when they are very small. Thus we have a system with no moving parts, passively generating power purely by the condensation of dew. At the moment it's just a laboratory experiment, but the authors think that it is scalable at low cost and provides another way to get renewable energy from the environment.

Miljkovic, N., Preston, D.J., Enright, R., Wang, E.N., 2014. Jumping-droplet electrostatic energy harvesting. Applied Physics Letters 105, 013111. http://bit.ly/1oNICeR

Wearable Energy-Generating Cloth May Replace Batteries

by Niti Nagar

Wearable electronics have been gaining traction among consumers through the decades beginning with light-up shoes and now evolving to smart watches and glasses. However, the major drawback of these gadgets' versatility is in short-lived batteries. In an attempt to overcome this limitation, scientists created a durable, flexible cloth that harnesses human motion to generate energy. Sang-Woo Kim and his team turned to the emerging technology of "triboelectric nanogenerators," or TNGs, which harvest energy from everyday motion. Kim and his team incorporated TNGs in a fabric of a silvery textile coated with nanorods and a silicon-based organic material. When the team stacked four pieces of this novel cloth together and pushed down on the material, it captured the energy generated from the pressure. The TNGs in the material immediately pumped out that energy which was used to power light-emitting diodes, such as a liquid crystal display or a vehicle's keyless entry remote. The cloth worked for more than 12,000 cycles, revealing high hopes for a sustainable and life-changing source of renewable energy.

In the American Chemical Society Journal, Nano, scientists reported it is the first of its kind that can also self-charge batteries or supercapacitors without an external power source. The energy generated from movement can be used to light up a small LED display, such as that on a wristwatch.

Kim and his colleagues indicate that the potential of sustainable wearable electronics goes far beyond convenience, flashiness or fashion, to include other types of new commercial products, as well as new medical applications. Biomedical technologies that require small power supplies, such as robotic skin, can use small, lightweight devices to drastically change the lives of many. However to maximize the utility and assert the feasibility of such technologies a more flexible and long-lasting energy source needs to be seamlessly incorporated into the device's design.

American Chemical Society. "Energy-generating cloth could replace batteries in wearable devices." ScienceDaily, 4 March 2015. <www.sciencedaily.com/releases/2015/03/150304110352.htm>.

Synthetic Diamonds Manipulated to be Sensitive Magnetic-Field Detectors
by Niti Nagar

MIT researchers have developed a new, ultrasensitive magnetic-field detector that is 1,000 times more energy-efficient than its predecessors. First published in April 2015 in Nature Physics, this technology relies on synthetic diamonds containing nitrogen vacancies (NVs). NVs are defects that are extremely sensitive to magnetic fields and capable of performing magnetic-field measurements. A diamond chip about one-twentieth the size of a thumbnail can contain trillions of NVs, which is the basis for the efficient and portable magnetometers. When hit with a laser light, the NV absorbs and re-emits the light, which carries information about the vacancy's magnetic state. Assistant Professor in Electrical Engineering and Computer Science and designer of the new device, Dirk Englund says, "In the past, only a small fraction of the pump light was used to excite a small fraction of the NVs. Now we make use of almost all the pump light to measure almost all of the NVs."

The electrons are often excited in NVs by directing laser light at the surface of the chip. To ensure most of the light is absorbed, researchers added a prism facet to the corner of the diamond and coupled the laser into the side. First author of the paper Hannah Clevenson says, "All of the light that we put into the diamond can be absorbed and is useful." Furthermore, they calculated the angle at which the laser beam should enter the crystal so that it will remain confined, bouncing off the sides in a pattern that spans the length and breadth of the crystal before all of its energy is absorbed.

Unlike a pure diamond composed of a carbon lattice, which doesn't interact with magnetic fields, a NV is a missing atom in the lattice, adjacent to a nitrogen atom. Electrons in the vacancy do interact with magnetic fields. As a photon strikes an electron in the NV, it jumps to a higher energy state. When the electron falls back down to its original state, the excess energy is released as a photon. A magnetic field can flip the electron's magnetic orientation, or spin, increasing the difference between its two energy states. Therefore the stronger the field, the more spins it will flip, ultimately changing the brightness of the light emitted by the vacancies. Making accurate measurements with this type of chip requires collecting as many of those photons as possible. Therefore it is important to maximize the amount of light absorbed by the chip.

The light can travel close to a meter within the chip. Englund describes it "as if you had a meter-long diamond sensor wrapped into a few millimeters." The geometry of the NVs is such that emitted photons emerge at four distinct angles. MIT researchers have placed a lens at one end of the crystal can collect 20% of them and focus them onto a light detector, which is enough to yield a reliable measurement. This also contributes to the efficiency of the device.

This technology could be used to make miniaturized, battery-powered devices for medical and materials imaging, contraband detection, and even geological exploration. Existing magnetometers are already used for these applications, however existing technologies have major drawbacks. Some rely on gas-filled chambers while others work only in narrow frequency bands thus limiting their utility.

Massachusetts Institute of Technology. "Better sensors for medical imaging, contraband detection." ScienceDaily. ScienceDaily, 6 April 2015. <www.sciencedaily.com/releases/2015/04/150406144606.htm>.

SolePower: Solving the Mobile Energy Problem
by Shannon O'Neill

Advancements in technology, specifically in handheld devices and portable electronics, are increasing at a rapid rate. Because battery technology and advancements have been moving at a much slower rate, the use of these devices has been limited to their battery life. This issue motivated engineering students from Carnegie Melon University to develop SolePower, a rechargeable battery that is powered and charged when the user walks.

A special insole (or "ensole", for energy insole) is placed in the user's shoe. The mechanism inside the insole is able to capture the kinetic energy produced when walking, which is then used to spin an electromagnetic generator as fast and as long as possible. The power created is then stored in an energy pack, which can be stored on top of the user's shoe or on the users ankle. This energy pack can then be hooked up to cell phones or other portable devices and used a portable battery. Currently, an hour walking provides enough energy to sustain two and a half hours of talk time on a cell phone, with a walk between two and a half to five miles providing a full charge to an iPhone.

Though this product is still under research and development in hopes of making it as compact and as efficient as possible, the company has over 600 backers, received numerous awards, and already has a waitlist of over five thousand people for when the product is released. Looking forward, the founders of SolePower are hoping to also create a low-cost model, specifically to be used in developing countries. This would be life changing for those who live in third world countries, as cell phones provide a crucial link in advancing. Specifically looking at Kenya as an example, 84% of the population owns cell phones while only 14% has access to electricity. Therefore, the ensoles provided by sole power would provide an energy source under any weather condition, as long as the user can walk. SolePower will also allow developing worlds to transition away from kerosene lighting, which is highly polluting and inefficient.

SolePower (http://solepowertech.com)

Smithsonian Magazine (http://www.smithsonianmag.com/innovation/generating-power-one-step-at-a-time-180953436/?no-ist)

Huffington Post (http://www.huffingtonpost.com/2013/09/06/solepower-mobile-charging_n_3882835.html?ncid=edlinkusaolp00000003)

BBC News (http://www.bbc.com/news/technology-23240968)

Inquisitr (http://www.inquisitr.com/692953/solepower-shoe-charge-your-phone-while-you-walk/)

Heat Pumps Placed in Frigid Bodies of Water can Nevertheless Offer Clean, Cheap Heating

by Trevor Smith

The town of Drammen, Norway, has been able to dramatically reduce its heating costs while increasing its heating supply through the use of heat pumps placed at the bottom of a nearby, frigid fjord. Since 2011, almost all of Drammen's heating needs have been met by the pumps, which function through a combination of temperature and pressure differences between the water in the fjord and ammonia in the heat pumps. In that time, the pumps have come to save the city approximately 2 million euros per year, in addition to reducing carbon emissions by 1.5 metric tons annually (Anderson 2015). Others are taking note; one Scottish company hopes to use a similar system to dramatically reduce Glasgow's heating costs (Bing 2014).

Although the pumps are placed in bodies of water with temperatures as low as 45 °F, these lakes and rivers are still warm enough to heat pressurized liquid ammonia until it evaporates. The ammonia gas is further pressurized, increasing its temperature to nearly 250 °F. This gas is used to heat water in the heating system, cooling the ammonia back down for it to begin the cycle again (Anderson 2015). The system could theoretically be installed next to any municipality near a body of water, standing or running.

One city considering this option is Glasgow, the largest city in Scotland. Because of its immense water system of rivers and lakes, Scotland is uniquely situated to take advantage of this opportunity. Star Renewable Energy, a clean energy company based out of Glasgow, is considering implementing heat pumps in both small and large scale projects. Star has proposed placing pumps in the River Kelvin to heat Glasgow University, claiming that it could cut the university's heating costs by 1.6 million pounds annually. But by installing a system of pumps all across Scotland, Star Renewable Energy suggests that Scotland could cut a massive 250 million pounds per year from its heating bill.

Anderson, Richard. "Heat pumps extract warmth from ice cold water". BBC News, March 9, 2015. http://www.bbc.com/news/business-31506073

Bing, Lemley. "Could Loch Ness be hiding monster energy savings?" businessGreen, April 25, 2014. http://www.businessgreen.com/bg/feature/2341378/could-loch-ness-be-hiding-monster-energy-savings

PHOTOVOLTAICS

The most rapidly evolving branch of renewable energy is photovoltaics. Unlike wind and hydropower which are basically electric fans and pumps run in reverse, and which haven't changed fundamentally for decades, maybe even centuries, photovoltaics are electronics and susceptible to something akin to Moore's law, getting exponentially more efficient and cheaper daily, and using less energy and materials to construct and install as well.

Gordon Moore, a founding member of Intel figured that the density of transistors on a chip would increase exponentially for the foreseeable future. In the case of photovoltaics, which are themselves made out of the same sort of stuff as transistors, the effect isn't packing more of them into the same space, but increasing the percentage of the energy in sunlight a panel can convert into electricity—their efficiency. This is, of course, limited to 100% of the energy in sunlight, but photovoltaics are nowhere near that yet and are rapidly improving just as the costs and impacts of manufacturing and installing them is decreasing. Furthermore, they are becoming flexible and suitable for incorporation into all sorts of devices and materials that seem counterintuitive—such as window glass and the threads that are woven into clothing. Some even convert the solar energy directly into hydrogen.

China and Kenya Partner to Promote Solar Energy Technology
by Mariah Valerie Barber

On April 25, 2015 investors from Kenya and China worked together to launch a solar energy transfer and training center. This center, China-Kenya Solid State Lighting Technology Transfer Center, will focus on transferring solar energy harnessed in Nairobi to other cities and towns throughout Kenya. The solar energy being produced at the transfer center will provide schools, households, businesses, and hospitals with affordable and reliable solar energy. The center will also focus on training Kenyans on how to operate and promote new solar energy technology systems. The center, along with the installment of a solar lighting system, is located in an industrial park in Nairobi.

The key investors and organizations behind the inter-state energy initiative is Sunyale Africa Limited, which dominates much of the solar energy field in Kenya and a company based in Beijing, China. Maina Maringa, director with Sunyale Africa Limited stated, "Our partnership with a Chinese firm will facilitate the establishment of a local assembly plant for solar products. The new center will expose Kenyan technicians to the latest solar solutions." Kenyan trainee and engineer, Richard Kipkorir stated that, "The Chinese are really helping us in terms of technology and information sharing. We think after some time doing this, we can gain a lot."

This China-Kenya partnership is representative of the spark in foreign investment into the renewable energy sector due to Kenya's policy and regulatory environment. The partnership allows for the Chinese firm to import solar products into the transfer center in Nairobi and then assemble them there once they arrive. The transfer center aims to spur more renewable energy growth in Kenya. The center will be especially beneficial for small businesses, factories, and entrepreneurs in Kenya, as they continue to demand more solar, renewable energy as a way to promote greater access to solar energy in remote, rural, or poor areas within Kenya.

PV Magazines: Photovoltaic Markets & Technology (http://www.pv-magazine.com/news/details/beitrag/china--kenya-establish-solar-technology-transfer-center_100019259/#axzz3YdnJp2GH).

Yibada. (http://en.yibada.com/articles/29527/20150427/china-kenya-partnership-solar-technology-green-energy.htm

Inspired by Nature: The Bionic Leaf
by Hannah Brown

Biomimetics is the principle of using processes found in the biological world and adapting them for specific, technological human needs. An example of this is nanotechnology. Inspired by the ways viruses operate, researchers have developed miniscule drugs that can target and treat specifically cancerous cells. We also use nature inspired products every day. Velcro, for example, was developed after a Swiss engineer studied the construction of tiny plant barbs that so easily stick to clothing. (science.com)

A new and exciting type of biomimetics comes in the form of the "bionic leaf." Reported on in a study released in the end of 2014 by Proceedings of the National Academy of Sciences, the bionic leaf was developed starting in 2009 by researchers now at Harvard. The idea is a form of reverse combustion, where CO_2 is turned back into fuel, and abiotic and biotic catalysts are used in conjunction to successfully transform energy into fuel.

It works like this: water is split using a catalyst made of earth metals, cobalt-phosphate, to make hydrogen. As hydrogen is rarely used as a transportation fuel or for electricity, a specifically engineered form of bacterium *Ralstonia eutropha* is used to convert the CO_2, H_2 and O_2 made from the water-splitting into isopropanol, a liquid form of energy.

Essentially, this bioreactor, powered by photovoltaic solar cells in this case, feeds starved microbes hydrogen from the water that has been split by the specific catalysis.

As microbes can turn energy into other molecules, the researchers introduced the specially engineered soil bacteria, *R. eutropha*, that uses the hydrogen as energy to build molecules out of carbon. From this process, isopropanol is created. While isopropanol is a form of liquid fuel, it is not as commonly used as others. However, the researchers believe that with adjustments and further research, the bionic leaf could produce other fuel, or pharmaceuticals, among other things, with just the combination of sunlight and CO_2 (salon.com).

As stated in the original study "liquid solar fuel derived from CO_2 holds promise as both a storage mechanism for solar energy, and as a renewable, carbon-neutral, and infrastructure-compatible energy supply." (Torella *et al.* 2014).

This bionic leaf is an intriguing form of biomimicry. As it combines non-living and living processes, and in the end creates liquid fuels from solar energy (a sort of electrofuel), it shows the variety of possibilities in the junction of biological, electrical and chemical engineering fields.

Joseph P. Torella, Christopher J. Gagliardi, Janice S. Chen, D. Kwabena Bediako, Brendan Colón, Jeffery C. Way, Pamela A. Silver, and Daniel G. Nocera. Efficient solar-to-fuels production from a hybrid microbial–water-splitting catalyst system. PNAS 2015 ; published ahead of print February 9, 2015, doi:10.1073/pnas.1424872112

Biello, David. "'We think we can do better than plants': New 'bionic leaf' makes fuel from sunlight." Salon.com. February 15, 2015

Strickland, Jonathan. "Top 5 Ways Nature Has Inspired Technology." Science.howstuffworks.com. April, 23, 2009.

Copenhagen, the New City of Lights

by Hannah Brown

Copenhagen has a lofty goal: to be the world's first carbon-neutral capital by 2025. Just 10 years away, it is implementing various technological tactics that are appealing for their cost-effectiveness, their ingenuity, and their capacity to save energy.

Thanks to an array of sensors embedded in the light fixtures that collect and feed data into software, using a wireless communication system, Copenhagen has already developed technology to help its citizens and encourage energy efficient transportation. Essentially, the city is using LEDs to create a sensory network that can coordinate a vast number of functions and services, some which have already been implemented.

One way in which Copenhagen is working towards its goal is through the introduction of the "green wave," green LED lights embedded in the ground to help bicyclists, riding in their own lanes, sync with their traffic lights, so that they don't have to stop and providing them the safest routes. They are also using LED streetlights that are motion sensitive and brighten only as vehicles get near, and turn down after they have passed. Additionally, there is a mobile tech side to this movement. For example, truck drivers now can use their smartphones to get updates on when the next light will change and bicyclists' movements are tracked through GPS to monitor traffic, and to give them the right of way if there are more than five cyclists at an intersection.

The city has even more ambitious plans for their remote sensing, including informing the sanitation department when the trash needs to be picked up. It will also assist in easing traffic congestion by specifically timing lights and dealing with the difficulties of weather problems, such as predicting where to salt before a snowstorm.

Using LED lights and sensors is not entirely a new idea. Los Angeles already uses sensors the detect traffic congestion and synchronize signals. Companies such as Cisco Systems and Sensity Systems work internationally, from Bangalore to Barcelona, coordinating these systems. But Copenhagen is using them on an even larger scale, and in conjunction with other technology such as solar powered streetlamps with small wind turbines on top and mobile technology that communicates with government systems to achieve goals, such as prioritizing the passage of bicyclists and buses over cars at intersections. Take the truck drivers as another example, while it may seem trivial that they have an app to time their route, it is actually very smart as stopping and starting the truck costs diesel fuel and money, which adds to noise and air pollution. This smart tech is attractive to city managers, as it will save money and energy, while improving the city's life in general. The LED sensors, and the data they collect, provide a technological perspective on encouraging energy efficient behavior and can be an example for cities across the world to follow. (nytimes.com)

Cardwell, Diane. "Copenhagen Lighting the Way to Greener, More Efficient Cities" nytimes.com, Dec. 8, 2014.

PosiGen is Making Solar Accessible
by Hannah Brown

As a burgeoning technology, photovoltaic solar cells have a reputation for being inaccessible to the general public or useful solely for those who have the expendable income to promote green energy technology. A Louisiana company, PosiGen, is trying to change that. (posigen.com)

Instead of being 'just another solar installation company', PosiGen sees itself as a solar-plus-efficiency enterprise. Called "blue-collar green" by its founders Thomas Neyhart and Aaron Dirks, PosiGen targets middle and low-income homeowners as it combines leasing low-cost solar with energy-efficiency upgrades. As 85% of Louisiana citizens are not permitted to go solar because they do not have the income or credit score to be part of traditional third-party leasing plans, PosiGen is opening up green tech to thousands of people.

What is unique about PosiGen is that it is not like other solar installation companies–it sells a whole speedy energy efficiency package. First things first, PosiGen installs the same system on every household it works with–no matter the size, no matter the location, it is always a 6-kilowatt system. In doing so, they drastically cut down on permit-approval time and have created an efficient assembly line technique for installing, which means that the job is started in less than 30 days, and there is a significant reduction in labor cost. In sacrificing customization, PosiGen claims to provide the greatest savings at the lowest costs.

So yes, they do do solar. But they also add up to 30 efficiency upgrades, such as swapping out incandescent light bulbs, adding insulation and putting in smart thermostats. Combined with the benefits of the solar panels, Neyhart estimates that customers save 40 to 80% more than when using a typical solar PPA.

All of the upgrades, solar and efficiency both, come to a final tally of $60 a month. The customers get a guaranteed percentage of the energy savings, such as $65 per month in Louisiana. As Neyhart says, "if you save someone who makes $150,000 a year $60 bucks a month, they're going to spend it on Starbucks, if you save that for a working family, they're going to spend it on their children, on school supplies, on groceries and on real needs for the home."

In the last three years alone, at a rate of working on 80 to 100 homes a week, PosiGen has set up more than 29 megawatts of solar on more than 6,000 homes across the country, primarily in low-income neighborhoods. PosiGen is so impactful that 60% of its customers come from referrals. A good product means positive reviews, and PosiGen seems to be using theirs well (greentechmedia.com).

Pyper, Julia. "PosiGen Brings Solar to the Working Class With a Unique Twist on a Lease." Greentechmedia.com. April 6, 2015
Posigen. http://www.posigen.com

From Sunshine Comes Potable Water
by Hannah Brown

When I was young, my best friend and I used to play on swing sets pretending that we were on a ship sailing the vast sea to save the world. Having a basic understanding that water was at the foundation of all life, we pretended that swinging on one of the swings made fresh water out of the sea saltwater. The other swing magically made gold.

Though this story is a silly illustration of a child's fantasy, desalination of water is an important key to solving drought problems across the globe. It is especially potentially important for farmers in developing nations.

To address this potential, USAID inspired researchers to compete for the Desal Prize, a competition to create affordable desalination solutions for developing countries. Of course, the system must remove salt from water, but it also has to meet three criteria. It must be cost-effective, environmentally sustainable, and energy efficient. The prize totaled $140,000, a substantial sum for innovation and research into the further utilization of energy efficient technology (Techxplore.com).

The winners of the first place prize were from MIT and Jain Irrigation Systems. For the energy efficiency and sustainability criterion, they designed a system that uses solar panels to charge a pack of batteries that powers the system that removes salt from water through the electrodialysis reversal method.

Electrodialysis removes the electrically charged sodium chloride ions in water that are the result of dissolved salt. They are taken out of the water by applying an electrical current, from two panels on each side, a cathode and an anode, which separates the ions from the water molecules as they are attracted and pulled with these opposite charges. This winning solution also applies UV light to the water to disinfect it (Wikipedia.com).

Using solar energy to desalinate water is not an entirely novel process. Solar desalination plants are already popping up in places like Chile and California, who that have been hit hard with droughts. What is unique, though, to systems for developing countries is the focus on durability. The winning team has tested their products in New Mexico, running the system for 24 hours straight, removing salt from more than 2,100 gallons of water each day. To ensure that their system is compatible with harsher climates, the MIT/Jain team will continue testing their system with rural farmers where USAID is already in place. The hope is that this system will be able to provide enough water to irrigate small farms, and encourage sustainable agriculture (Popsci.com).

Greggs, Mary Beth. "MIT Invention Turns Salt Water Into Drinking Water Using Solar Power." April 23, 2015. Popsci.com

Yirka, Bob. "MIT and JAIN team wins the Desal Prize for desalination system" April 27, 2015. Techxplore.com

Electrodialysis. Wikipedia.com

iGrenEnergi Develops Technology to Boost Productivity of Solar Panels

by Nour Bundogji

iGrenEnergi, a startup headquartered in Mumbai, invented a technology that could give solar energy "a whole new ray of hope" as Shonali Advani claims in his recent post in the Economic Times. With their eight-panel DC Optimizer (DCO), iGrenEnergi can maximize a solar panel's energy capacity through its life cycles.

Advani reports, "It addresses a pertinent problem – blockages to sunlight such as shading or any particle like dust, bird droppings, or leaves that inhibit energy falling on a panel."

Present day solar technology only allows for 19–22% of light hitting a panel to get converted into electricity. Co-founder of iGrenEnergi, Sunit Tyagi, says, "One shaded panel kills production of all panels. So, if 90% of one panel is not blocked, each panel loses 10% in the string and one guy can impact 20 panels. Our DCO addresses this."

This patented technology can be placed next to a solar panel and can save 10-40% of energy.

How? iGrenEnergi hasn't released the exact mechanism of their patented DC Optimizer. But, overall it's a device that uses an algorithm to tell the panels to work at what capacity, referred to as 'energy packetization.' Most DC optimizers operate by individually tuning the performance of a solar panel through maximum power point tracking (MPPT), which helps tune and get the maximum power from a photovoltaic module. To ultimately tune the output to match the performance of the string inverter—which converts the variable direct current output of a photovoltaic solar panel into a utility frequency alternating current that can be fed into a commercial electrical grid or used by a local or commercial electrical grid (MacAlpine *et al.* 2012).

"The potential of the product rests on these strategic pillars—reusable hardware platform, a cloud based application software to analyze data, and product-specific embedded software to control energy packets flowing through the hardware," reports Advani.

Although the device is in the pilot stage, it has been installed in two locations, a residential rooftop and a small manufacturing company, to test for effectiveness. At the residential location, the DCO showed a 10–12% increase in energy efficiency.

"This product seems promising, but it's a matter of time to see how many people will go for a solar installation just because of the DCO. At the urban level, installations are going up but the pace is slow. The basic opportunity for this depends on how fast solar installations go up. However, since it's a startup, if they can sell to existing installations, it can be a good revenue earner in initial few years," said Pamli Deka, consultant, Regin Paradise Consulting that runs New Ventures, a clean technology innovative center supporting sustainable energy development.

Regardless, the company has secured investment from various individuals from Europe and the US that sum up to $0.5 million and is about to close another round of $0.5 million.

Advani, S. iGrenEnergi working on 'energy packetization' technology to boost productivity of solar panels. The Economic Times. March 8, 2015
http://economictimes.indiatimes.com/news/emerging-businesses/startups/igrenenergi-working-on-energy-packetization-technology-to-boost-productivity-of-solar-panels/articleshow/46498587.cms
MacAlpin, S.M., Characterization of Power Optimizer Potential to Increase Energy Capture in Photovoltaic Systems Operating Under Nonuniform Conditions. IEEE Transactions of Power Electronics. 28,6, June 2013
http://ieeexplore.ieee.org/stamp/stamp.jsp?tp=&arnumber=6340351

A Fully Transparent Photovoltaic Cell

by Nour Bundogji

Imagine a world where every window and smartphone screen is tuned into a solar cell. Sounds like a self-sustainable world, right?

Researchers at Michigan State University have created a fully transparent solar concentrator turning every sheet of glass into a photovoltaic solar cell. Unlike previous attempts at such a device, Michigan State University's solar cell is completely transparent.

Usually photovoltaic cells make energy by absorbing photons and converting them into electricity. However, if the material is transparent, by definition, all of the incident light passes through it. Thus, previous transparent solar cells have only been partially transparent and cast a shadow of color that can be quite distracting.

The technology that the Michigan State researchers use is slightly different. Instead of creating a transparent photovoltaic cell, they use a transparent luminescent solar concentrator (TLSC). The TLSC consists of organic salts that absorb specific non-visible wavelengths of ultraviolet and infrared light, which allows them to glow as another wavelength of infrared light. This emitted infrared light is then guided to the edge of the plastic, where thin strips of conventional photovoltaic solar cells convert it into electricity (Zhao *et al.*, 2014).

While non-transparent luminescent concentrators have a maximum efficiency at around 7%, TLSC currently has an efficiency of only around 1%. However, the authors believe they can increase this efficiency to 5%. Although these figures aren't large on their own, they could quickly add up if every window of a house or office had TLSCs installed.

The researchers are confident that this technology can not only be affordable, but also can be scaled from large industrial and commercial applications to consumer devices. According to Richard Lunt, who led the research team, "The team is confident that the transparent solar panels can be efficiently deployed in a wide range of settings, from tall buildings with lots of windows to any kind of mobile device that demands high aesthetic quality like a phone or e-reader."

Anthony, S. A fully transparent solar cell that could make every window and screen a power source. August 26, 2014. http://www.extremetech.com/extreme/188667-a-fully-transparent-solar-cell-that-could-make-every-window-and-screen-a-power-source

Zhao, Y., Meek, G., Levine, G., Lunt, R. (2014) Near-Infrared Harvesting Transparent Luminescent Solar Concentrators. Advanced Optical Material. DOI: 10.1002/adom.201400103

A Self-Powered Video Camera that Can Run Indefinitely without an External Power Supply
by Nour Bundogji

A research team at Columbia Engineering invented a prototype of the first self-powered video camera. With this prototype the team of researchers, led by Shree K. Nayer, were able to indefinitely produce an image each second of a well-lit indoor scene. How?

Early in the development process, Nayer realized that "although digital cameras and solar panels have different purposes—one measures light while the other converts light to power—both are constructed from the same components." For instance, at the heart of any digital camera is an image sensor made of millions of pixels. More specifically, the photodiode of a pixel is what produces and extracts current when exposed to light, enabling each pixel to measure the intensity of light falling on it. A similar photodiode found in solar panels allows incident light to be converted to electric power. What sets the two photodiodes apart is the one in a digital camera is in a "photoconductive mode" while in a solar cell it is in a "photovoltaic mode." Thus, Nayer worked with an undergraduate research engineer, Daniel Sims BS'14, and consultant Mikhail Fridberg of ADSP Consulting to merge these two modes together in their prototype video camera. In other words, they designed a pixel that can not only measure incident light but also convert the incident light into electric power.

A commentator at ScienceDaily.com briefly explained the technology, "During each image capture cycle, the pixels are used first to record and read out the image and then to harvest energy and charge the sensor's power supply, continuously toggling between image capture and power harvesting modes. When the camera is not used to capture images, it can be used to generate power for other devices, such as a phone or a watch."

Nayar continued, "A few different designs for image sensors that can harvest energy have been proposed in the past. However, our prototype is the first demonstration of a fully self-powered video camera." He continues "And, even though we've used off-the-shelf components to demonstrate our design, our sensor architecture easily lends itself to a compact solid-state imaging chip. We believe our results are a significant step forward in developing an entirely new generation of cameras that can function for a very long duration—ideally, forever—without being externally powered."

It is their hope that this technology can infiltrate different wearable devices sensor networks, smart environments, personalized medicine, etc.

Columbia University School of Engineering and Applied Science. "A video camera that powers itself." ScienceDaily.com. ScienceDaily, 15 April 2015. <www.sciencedaily.com/releases/2015/04/150415102924.htm>.

Collaboration for Efficient and Reliable Solar Tracking Technology for Commercial Businesses
by Jessie Capper

Solar tracking technology has proven to improve the performance of photovoltaic panels by 20% or more over fixed systems. Dual-axis tracking systems—moving along both the x and y axes—typically provide 8 to 10% more energy than single-axis systems. Although solar tracking technology improves performance of photovoltaic panels, there are concerns regarding the land requirements and high costs associated with this innovative technology (Renewable Energy World June 13, 2013). Trackers tend to be more expensive due to their moving parts; therefore, the motor needs to be professionally installed and properly maintained. Furthermore, tracking systems' movement can cast shadows on neighboring panels; as a result, these systems require more land than fixed solar systems. Due to their advanced technology and demand for land space, solar tracking technology can be a costly investment.

Fortunately, SunPower Corp, AllEarth Renewables, and SolarSense have collaborated to maximize energy production for large-scale solar projects with solar tracking technology for photovoltaic panels (Renewable Energy World Jan 19, 2015). These companies have focused on advancing the dual-axis tracking technology at a low cost while maintaining reliability and efficiency. AllSun Solar Trackers, manufactured by AllEarth Renewables in Williston Vermont, use a motor and GPS to rotate the solar PV panels all directions—from east to west, to up and down—to provide the maximum energy production possible; these trackers result in 45% more energy than rooftop solar (AllEarth Renewables). SunPower Corp. has equipped these trackers with SunPower solar panels; SunPower's panels have been demonstrated to be the most efficient solar photovoltaic panels commercially available in the industry (SunPower). The success in the combination of these two technologies, and the collaboration among these three businesses has built a promising platform for the future of solar tracking technology for renewable energy.

"SolarSense Completes Development and Financing of 2.15 MW of AllEarth Solar Tracker Projects." Renewable Energy World. January 19, 2015. Accessed February 11, 2015. http://www.renewableenergyworld.com/rea/companies/allearth-renewables/news/article/2015/01/solarsense-completes-development-and-financing-of-2-15-mw-of-allearth-solar-tracker-projects

"Solar Tracking Systems Gain Ground." Renewable Energy World. June 13, 2013. Accessed February 15, 2015. http://www.renewableenergyworld.com/rea/news/article/2013/06/on-track-to-succeed-with-solar-tracking-systems

AllEarth Renewables (http://www.allearthrenewables.com/our-product/).
SunPower (http://us.sunpower.com/).

Pursuing the Next Frontier for Solar Technology: Northrop Grumman and Caltech's Space Solar Initiative
by Jessie Capper

Despite the decreasing cost of solar and wind energy here on Earth, the Northrop Grumman aerospace company, in collaboration with the California Institute of Technology (Caltech), recently announced plans for their Space Solar Initiative (Casey April 22, 2015). The Space Solar Initiative aims to develop a space-based solar array that can generate electricity as inexpensively as fossil fuels (Space Solar Initiative). Although this would be beneficial for the advancement of renewable energy, there are some doubts that the venture is worth the $17.5 million recently provided by Northrop Grumman, especially when considering the diminishing costs of renewable energy on Planet Earth (Casey April 22, 2015).

These concerns, however, do not seem to be preventing the progress made on space-based solar power as the technology continues to demonstrate several benefits over terrestrial solar and wind farms. Real estate is a primary example of this. Many solar farms take advantage of rooftops, parking lots, brownfields, and other pre-developed sites; however, many solar farms are built on valuable real estate. According to CleanTechnica writer Tina Casey, this real estate includes former farmland, which has, in turn, engendered tension with "agriculture and habitat conservation" (Casey April 22, 2015). Wind turbines have a bit more flexibility when it comes to an open space footprint, but their large size also limits site options (Casey April 22, 2015).

When considering these drawbacks to terrestrial renewable energy, it is apparent that space solar power has many advantages. Other benefits of space solar power focus on the relative ease of transmitting the energy from space to earth. Solar development on Earth is best in desert regions and offshore ocean sites; however, these locations require long transmission lines which are expensive and present multiple risks with respect to vulnerability to natural disasters and human intervention (Casey April 22, 2015). Lastly, according to research from the National Space Society, a network of space-based solar power arrays provide a reliable source of solar power independent of time of day or weather on Earth; this thereby reduces the need for costly energy storage facilities (National Space Society).

Despite the many advantages of space solar power, the new Northrop Grumman Space Solar Power Initiative at Caltech must anticipate three challenges to achieving their goal. The first challenge is the weight of the photovoltaic cells (PV cells); the PV cells must be sufficiently lightweight to be properly deployed while also being ultra-efficient (Casey April 22, 2015). Similarly, the second issue is the weight of the structure on which the solar cells will be arrayed. Lastly, Northrop Grumman and Caltech must resolve how the system will transmit the solar generated electricity down to Earth from space. These three problems do not seem to be of much concern to Caltech or Northrop Grumman however. According to researchers at the Naval Research Laboratory, the US Navy space solar power project, demonstrated that the transmission of solar generated electricity from space to earth is relatively simple (Casey March 13, 2014). According to Dr. Massimiliano Vasile at the University of Strathclyde Institute for Energy and Environment, microwaves or lasers are potential technologies that will enable solar energy captured in space to be transmitted directly to specific areas on earth (PlanetSave, May 17 2012). These technologies "would also avoid having to address the vexing problem of storing intermittent energy from renewable sources as it would provide a constant stream of energy" (PlanetSave, May 17 2012). In addition, Caltech and Northrop Grumman have a successful history of collaboration dating back to the 1930s. Furthermore, Caltech is currently managing the Jet Propulsion Laboratory for NASA, thereby gaining expertise and knowledge on how to best address these aforementioned issues. Due to the positive perspective and progressive momentum of the two companies, the plan is to build the Space Solar Power Initiative up to include 50 researchers at Caltech who will work

with Northrop Grumman's team to begin building prototypes
and successfully achieve their mission (Casey April 22, 2015).

Casey, Tina. "Race For Space Solar Power Heats Up With Northrop Grumman, Caltech
    Partnership." CleanTechnica. April 22, 2015. Accessed April 22, 2015.
    http://cleantechnica.com/2015/04/22/race-space-solar-power-heats-northrop-
    grumman-caltech-partnership/
Casey, Tina. "Yes, Space Solar Power Is A Real Thing." CleanTechnica. March 13, 2014.
    Accessed April 23, 2015. http://cleantechnica.com/2014/03/13/us-navy-
    develops-space-solar-power/
National Space Society (http://www.nss.org/settlement/ssp/)
"Space-based Solar Power Systems a Step Closer to Reality." PlanetSave. May 17, 2012.
    Accessed April 28, 2015. http://planetsave.com/2012/05/17/space-based-solar-
    power-systems-a-step-closer-to-reality/
Space Solar Initiative
    (http://www.globenewswire.com/newsarchive/noc/press/pages/news_releases.h
    tml?d=10129649)

Solar Power in Space
by Alex Elder

Although solar power is currently a popular source of
renewable energy in the United States, it has several drawbacks.
The biggest problem with relying on sunlight for energy is the
disruption of solar absorption when there is a lack of sunlight.
Thus, in areas of the country that often experience cloudy or
overcast weather, solar power is not a reliable source of energy.
Furthermore, solar power is not typically available at night
without a sophisticated energy storage system.

One solution to this problem is to establish solar farms in
space. Although it may sound like science fiction, colonies of
solar panel satellites in space would circumvent the major
disadvantages associated with using solar panels on Earth.
Additionally, sunlight in space is about ten times more powerful
than what we experience on Earth, making this system more
efficient in both its reliability and its strength.

In order to transmit the energy from space back to Earth,
an orbiting solar farm would need to absorb energy from the sun
and then convert it into radio waves. In this form, the energy
could be "beamed" down to receiving antennae on Earth. The
radio transmissions would then be converted back into useable
electricity and fed into the conventional power grid. The concept
of transmitting solar energy via radio waves has already been
tested in 2008 when NASA was able to beam solar energy
between two Hawaiian islands 90 miles apart (Iannotta, 2008).
However, because the receivers in the experiment were so small,
very little of the energy was actually received.

Although this proof-of-concept experiment was not entirely
successful, it opened up the doorway to future experiments in

space-based solar power and further development of wireless power transmission. Continued innovation in the energy sector will help generate new ideas for future sources of energy and methods of improving our current energy systems.

Iannotta, B. (2008). Experiment boosts hopes for space solar power. NBC News. http://www.nbcnews.com/id/26678942/#.VOzxb_nF98E

McSpadden, J. O., & Mankins, J. C. (2002). Space solar power programs and microwave wireless power transmission technology. Microwave Magazine, IEEE, 3(4), 46-57.

Reed, K., & Willenberg, H. J. (2009). Early commercial demonstration of space solar power using ultra-lightweight arrays. Acta Astronautica, 65(9), 1250-1260.

An Off-Grid Renewable Energy Generator
by Alex Elder

Although solar energy is an appealing form of clean and renewable energy, not all areas of the world have the technology or the infrastructure to utilize it. Most modern solar panel installations require a connection to the existing energy grid in order to function properly. Many remote communities, however, do not have this grid infrastructure in place and thus cannot use modern solar technology at a large scale.

SunEdison, a major player in the energy world, recently developed a new piece of technology called the Outdoor Microstation which aims to make renewable energy in rural and off-grid areas sustainable. The Microstation is a standalone power generator that can provide a renewable energy source to remote areas of the world. The unit includes photovoltaic solar panels to harness clean and renewable energy from the sun as well as a battery system to store unused energy. The 3,500-volt ampere model can power a rural community of 25 homes for 5 hours each night, including street lighting (SunEdison 2014). It is even scalable and modular, which provides consumers the ability to connect several systems. All Microstations are monitored remotely by SunEdision. The unit can also be installed quickly (in 4 to 6 hours) and is very low maintenance. The Microstation is also extremely durable and able to withstand extreme weather conditions, with a lifetime estimation of over 10 years. The low maintenance and durability aspects of the Microstation are just as important as its ability to generate power since a quick deterioration of the technology would ultimately fail to provide the promised energy source.

The development of the Outdoor Microstation is part of SunEdison's campaign called the Eradication of Darkness. This movement aims to bring electricity to 20 million people by 2020, with the interim goal of lighting up 1 million homes in 2015 (Tweed 2015).

SunEdison (2014). SunEdison Outdoor Microstation Datasheet.
http://www.sunedisonemea.com/docs/SunEdison-Outdoor-Microstation-Datasheet.pdf
Tweed, K. (2015). SunEdison's Next Market: Solar Minigrids and Micropower Stations for the Energy Poor. GreentechSolar.
http://www.greentechmedia.com/articles/read/sunedison-will-bring-electricity-to-20-million-by-2020

Jim Ayala and Hybrid Social Solutions: Innovation in Distributing Solar

by Liza Farr

In the past two years, Jim Ayala, Founder and CEO of Hybrid Social Solutions (HSS), has come to the forefront of entrepreneurs in solar technology. His company is a social business, with the mission to "Develop practical applications for existing technologies by understanding localized conditions and co-developing new product lines with customers" (World Economic Forum). Specifically, HSS works in the Philippines to provide solar-powered electricity access for the many remote, impoverished villages, 25% of which do not currently have access (World Economic Forum. Ayala gathered a solar energy network to do the negotiating with solar supplies to tailor products to local needs. The company's work has increased household cash flow by 25%, and improved health and safety conditions by eliminating kerosene fumes, fires, and accidental ingestion (World Economic Forum). Children are able to study 45% longer, and 97% feel safer (Energyboardroom, Jan 30, 2014). Using solar to power off-grid communities is not new, but HSS's personalized method of distribution and technology will change the way solar energy is pursued in developing nations.

The technologies Ayala and his company have come up with are catered specifically to the various needs of villagers, which they learn by visiting the villages themselves. They introduced solar-powered spotlights that shine 50 meters away, scaring away pests and reducing crop loss up to 30%. Fisherman often use kerosene flares to attract fish at night, which takes up 40% of the revenue. HSS designed a solar light that would attract the fish, saving the fishermen money and increasing health and safety (World Economic Forum). Ayala's most recent venture is through Siftung Solarenergie Foundation, where they are partnering with the Department of Education to implement solar libraries in classrooms across the County (Official Gazette, Aug

19, 2014). Students can check out solar reading lights to study at night, increasing study time for children, but also helping their families who can benefit from the light as well. Ten schools had received a solar library in mid 2014, and many more are being implanted, increasing safety of students and their potential for economic advancement (Official Gazette, Aug 19, 2014). Ayala sees these solar innovations as akin to cell phones. Many people in developing nations have cell phones because the areas do not have the infrastructure for landlines. Solar allows these people the same opportunities as those in less remote areas, without the high costs of electricity infrastructure (Energyboardroom, Jan 30, 2014). Because of their strategy of working with their customers to develop a truly useful product, and continuing education and maintenance of the technologies, HSS has been extremely successful with not only turning a profit, but also providing a lasting positive benefit to Filipino communities.

Energyboardroom. "Interview: Jim Ayala, Founder & CEO, Hybrid Social Solutions, Philippines." Jan 31, 2014.
[http://www.energyboardroom.com/interviews/interview-jim-ayala-founder-ceo-hybrid-social-solutions-philippines]
Official Gazette. "DepEd: Light for Education Program to provide solar libraries for rural schools." Aug 19, 2014. [http://www.gov.ph/2014/08/19/deped-light-for-education-program-to-provide-for-solar-libraries-in-rural-areas/]
World Economic Forum: [http://reports.weforum.org/social-innovation-2013/hybrid-social-solutions-inc-hssi/]

Solar Fabric is the Second Generation of Solar Technology
by Liza Farr

After seven years of research and development, Perry Carroll's Solar Cloth Company is putting its lightweight, flexible solar panels on the market (Solar Cloth Company). While sailing his yacht in the Atlantic Ocean, he was inspired to combine solar energy and fabric to enable solar power to cover more types of structures (Hickey, Mar 22, 2015). The new type of panel can be rolled and fitted on curved structures, as well as roofs that are not able to sustain the weight of glass panels (Hickey, Mar 22, 2015). The thin film photovoltaic is being called the second generation of solar technology (Hickey, Mar 22, 2015). The panels are 20% of the weight of standard panels, but also produce 15% less power and cost twice as much (Hickey, Mar 22, 2015). One parking lot cover, for example, costs $19,000 (Hickey, Mar 22, 2015). Perry assesses the economic viability of the product based on opening new markets with new siting possibilities for solar panels.

The company is marketing these panels for non-load bearing roofs and car parking structures, as well as for data centers, super markets, and warehouses (Solar Cloth Company). According to the Solar Cloth Company, there are 834 million square meters of non-load bearing roofing and 353 million square meters of car parking in the United Kingdom alone. These two potential markets are valued at $250 billion and $100 billion respectively (Solar Cloth Company). Much of the United Kingdom factories are also potential sites for the solar rolls, and they account for 13% of national energy consumption, making these panels a way to significantly reduce carbon emissions (Hickey, Mar 22, 2015). The company has already received over $1 million in orders (Hickey, Mar 22, 2015). However, investors were hesitant to fund the new technology, so the company crowd-funded $1.5 million (Hickey, Mar 22, 2015). Perry is a strong advocate for more research and development into solar energy in general in the United Kingdom (Hickey, Mar 22, 2015). Additional funding for his own research and development, as well as rising electricity prices will likely make the product more successful moving forward. If all else fails, Perry has a backup plan to make solar underpants. He made one pair for a Japanese businessman who gave them to his boss with the note "I told you the sun shone out of my backside" (Burn-Callander, Dec 6 2014).

Burn-Vallander, Rebecca. New solar 'cloth' to turn UK rooftops into batteries. December 6, 2014.
[http://www.telegraph.co.uk/finance/businessclub/technology/11274131/New-solar-cloth-to-turn-UK-rooftops-into-batteries.html].
Hickey, Shane. Solar Sails Set Course for a New Journey into Renewable Energy. March 22, 2015. [http://www.theguardian.com/business/2015/mar/22/solar-sails-set-course-for-a-new-journey-into-renewable-energy]
The Solar Cloth Company [http://www.thesolarclothcompany.com]

Solé Power from MiaSolé
by Alexander Flores

MiaSolé has advanced the solar panel industry through the development of lightweight, flexible, and power solar cells. These solar cells are composed of copper indium gallium selenide (CIGS), a tetrahedrally bonded semiconductor with a chalcopyrite crystal structure, which has the ability to be deposited on flexible substrate materials. The CIGS film is deposited on a thin stainless steel sheet and then cut into cell form with an ultrawire interconnect. The thin film cells are the highest in energy conversion efficiency and have the ability to be modified to fit any form factor or structure. This allows for the production of solar panels that are lighter, cheaper to install, and more durable. Panel costs are also reduced due to the higher conversion efficiency and the use of fewer panels, which results in less wiring, labor, and land costs. These panels are ideal for use on metal and low-slope commercial roofs, including thermoplastic polyolefin (TPO) single-ply roofing. Since 2010, MiaSolé has been able to increase panel efficiency from 10% to 15.5% and intend to overtake poly-Si panels within the next year. It seems like the solar panel industry should have its money on MiaSolé.

MiaSolé 1 (http://miasole.com/en/home/)
MiaSolé 2 (http://miasole.com/en/miasole-advantages/efficiency-leadership/)
MiaSolé 3 (http://miasole.com/en/miasole-advantages/flexible-solar-cell/)
MiaSolé 4 (http://miasole.com/en/product/modules0/)
MiaSolé 5 (http://miasole.com/en/references/lightweight-commercial-rooftop/)
MiaSolé 6 (http://miasole.com/en/references/residential-rooftop/)

New Solar Farm in China

by Dylan Goodman

Over the past several years China has become increasingly reliant on renewable energy sources. In the next 15 years China is expected to have more low-carbon energy than the capacity of the entire United States power grid (Randall). In making the switch to renewables, China is decreasing their reliance on fossil fuels. Just recently, Apple announced their plans to construct 40-megawatts of solar farm in Sichuan Province in China. This past February Apple announced plans to build a large solar farm in California, and now they're making a similar move internationally. Teaming up with SunPower Technologies, a California based solar company, Apple plans to construct and connect two 20-megawatt plants to the grid in China by the end of 2015. The plant is designed to utilize rows of mirrors reflecting light onto high-efficiency SunPower-designed solar cells.

Reducing fossil fuel use in China can be seen as a large environmental advantage. Many large US based retailers have been outsourcing manufacturing to China for a many years. Inevitably, this has led to increased pollution and decreased air quality. Apple's push away from fossil fuels could influence other companies to do the same. This project however marks a relatively small project for China's energy production as a whole. While 40-megawatts is a considerable amount of power, and one of the larger corporate-backed projects, China is expected to bring 41-megawatts online daily through the end of 2015 (Reardon). By making a push towards increased solar farms, apple is reducing carbon emissions while helping influence other retail corporations to do the same. "These projects will provide clean, renewable energy, help address climate change, and continue to provide agricultural benefits to the local farmers, while protecting the area's precious land." (Tom Werner, President and CEO, SunPower). The move toward increased renewable energy production is expected to have large long-term environmental benefits.

Randall, Tom. "Here's Why Apple Is Building Solar Farms in China." Bloomberg Business. N.p., 21 Apr. 2015. Web.
"SunPower to Partner with Two Solar Projects in Sichuan Province." RenewablesBiz. N.p., 20 Apr. 2015. Web

Perovskite Solar Cells

by Dylan Goodman

Recent research from Stanford University has produced a promising new outlook on Solar Panel efficiency. Perovskite, a mineral composed of calcium titanate, has been found to increase the efficiency of conventional solar cells. Functionally, solar cells work by converting light energy, in the form of photons, into electrical energy. With the future of energy uncertain, this extremely unique process allows humans to harness energy from the sun, a non-diminishing resource. If the technology to exist to harness the suns energy with perfect efficiency, we could easily power the entire planet.

Conventional solar cells, constructed mostly of silicon materials, absorb and convert both the visible spectrum as well as infrared light. However, this process is not always extremely efficient. While the non-visible light spectrum does contain energy, it is relatively low compared to the energy of visible light. Researchers hope to use perovskite to increase the overall efficiency of conventional solar cell technology. According to research by Stanford University scientists, by laying perovskites on top of existing silicon solar cells, they've been able to increase the efficiency of energy conversion by up to 50%. The process works by layering multiple solar cells on top of each other. While the bottom of such panels will consist of conventional silicon solar cells, the upper layer will be made of perovskites. Light passes through each level of the multilayer cells, and the silicone and perovskite are able to absorb different parts of the spectrum with increased efficiency. In doing so, the use of multiple materials allows for layered cells to convert substantially more light into energy. With the proper combination, efficiency can be improved by up to 50%.

Unfortunately, the long-term stability of perovskites is largely unknown. While silicone is a rock and can be heated to high temperatures with no risk, perovskite will degrade if exposed to light or water. While the technology is not ready to produce perovskite cells designed to last the 25 years of a conventional panel, research looks promising in developing a more durable cell in the next 5–10 years.

Bailie, Colin D. *et al.* 'Semi-Transparent Perovskite Solar Cells For Tandems With Silicon And CIGS'. Energy Environ. Sci. 8.3 (2015): 956-963. Web. 24 Mar. 2015.

Solar Power Duo: A Perovskite and Silicone Based Semiconductor
by Alison Kibe

A 2013 article for Nature by Michael McGehee puts forth that perovskites ($CH_3NH_3PbI_3$)–a family of semiconductor crystals— would quickly change the world of photovoltaics with their cheap and simple design. Solar cells produced for commercial use typically contain silicone semiconductors that can easily incur defects during the production process that cause efficiency losses over time and reduce the life span of a solar panel. However, they have shown the highest rates of efficiency (17–23%) compared to other potential semiconductor materials—until now.

As of 2013, perovskites showed great promise, with researchers producing solar cells that met a 15% efficiency benchmark. Since then, perovskite production methods have improved and perovskite-based cells have only become more efficient. Less than two years later, Robert Service has written an article for Science Magazine on how researchers have discovered the power of using perovskites in conjunction with silicone semiconductors. These special types of semiconductors, termed as "tandems" for their combinatorial use of perovskites and silicone, are the latest innovation in solar voltaic cell technology. Man-Gyu Park, from Sungkynkwan University in Suwon, South Korea, was able to use them to produce cells with an efficiency rate of up to 28%.

Reaching such a high efficiency point was not easy, so tandems are not yet ready for commercial use. The most efficient tandems work by using an optical splitter to divide high and low energy light and direct each type into the perovskite cell and silicone cell respectively. But what is remarkable about this improvement and has been noted before by those involved with perovskite research, is not just greater efficiency, but the speed at which development has occurred. Even today, researchers are already aware of the next steps they can take to improve efficiency and are finding better ways to produce perovskites. With more promising electricity storage technologies like liquid batteries (Energy Vulture.com 1) or the use of car batteries for storage (Energy Vulture.com 2), the use of renewable energy sources like solar may become more feasible sooner rather than later.

Service, R., 2015. Devices team up to boost solar power. Science Magazine 347, 225.
McGehee, M., 2013. Fast-track solar cells. Nature. 501, 323-324.
Energy Vulture.com 1: Liquid Batteries
    (http://energyvulture.com/2014/09/27/membrane-free-lithiumpolysulfide-semi-liquid-battery-for-large-scale-energy-storage/)
Energy Vulture.com 2: Car Batteries for Storage
    (http://energyvulture.com/2011/05/23/the-economics-of-using-plug-in-hybrid-electric-vehicle-battery-packs-for-grid-storage/)

Material Architecture: Graphene and Carbon Nanotube Applications for Energy
by Alison Kibe

With the availability of cost effective and easily scalable synthesis methods, researchers have begun working with porous and 3D graphene and carbon nanotube (CNT) structures. Wang, Sun, and Chen (2014) wrote a review article outline uses for foam-like structures of CNTs, graphene, and hybrids of the two. Using a process called chemical vapor deposition, it is possible to construct defect free 3D architectures. This type of method is currently used in thin film production, i.e. production of semiconductor wafers in photovoltaic cells.

What makes the porous 3D structure so useful is it's greater surface area and structural strength. Graphene is composed of an atom thick layer of carbon rings that gives it a stable structure but means that the edges of the layer are free to interact with other atoms and molecules. The surface area of porous 3D structures means that there are more reactive edges. This is ideal for the addition of chemical functional groups, biomolecules, polymers, and other nanomaterials.

One of the major potential applications for CNT/graphene hybrids is their ability to transport ions, fuel molecules, and electrons quickly. Issues found in biofuel cells like the difficulty of electron transfers between enzymes and catalysts could be alleviated via "decorating" a 3D graphene structure with enzymatic structures.

There is a hope for graphene and carbon nanotubes as transformative super materials that will provide a variety of energy applications. But, when will we actually see any of this technology in use? In an article for the New Yorker, John Colapinto asks the question, what is graphene actually for? He points out that other technologies, like the science behind magnetic resonance imaging (MRI) technology wasn't realized as the medical MRI machines we recognize today for nearly thirty years. In addition, in many cases graphene needs to be altered in order to be useful, but changing graphene can cancel out advantages it has over other materials. For this reason, even though graphene is known for its conductive properties, it is of little use as a semiconductor because semiconductors must be able to be turned on and off. Without changing graphene's conductive capacity, it is impossible to turn graphene off. Perhaps we will see more commercially applicable uses for graphene and CNT in the near future, but if history is any indication, it's a long way off. Still, it is probably only a matter of time.

Wang, X., Sun, G., Chen, P., 2014. Three-dimensional porous architectures of carbon nanotubes and graphene sheets for energy applications. Frontiers in Energy Research 2, 1–8.

Colapinto, John. "Material Question." The New Yorker. December 22, 2014. http://www.newyorker.com/magazine/2014/12/22/material-question

Expanding the Frontiers of Energy: Pay-as-You-Go Energy

by Alison Kibe

With little to no access to electricity grids in rural areas of Africa, the Nairobi based startup M-KOPA Solar launched in 2012 as an effort to provide affordable solar energy units to households in Kenya, Tanzania, and Uganda. A recent press release announced that M-KOPA is entering its fourth round of investment worth $12.45 million (Jackson, 2015). The money will be used to add products to M-KOPA's line, expand business into East Africa, and license their products for use in other markets (Jackson, 2015). The start up also won the Zayed Future Energy prize in February. Worth $1.5 million, the money will be used to start a development program called M-KOPA University that will focus on developing employees' business and technical skills (Mutegi, 2015).

The product M-KOPA's website (M-KOPA.com) currently offers is a home solar energy unit. Households can purchase the unit under a pay-per-use installment plan that costs $35 up front and an additional forty-five US cents per usage credit. After a year of payments, customers own the solar cells and can increase the amount they spend on power. This ends up costing less than kerosene lighting and does not create a fire hazard or emit harmful fumes.

The M-KOPA website also claims to add 500 new customers to its existing 150,000 customers every day. Part of what makes M-KOPA so effective is that everything is paid through existing cellphone service infrastructure. Cellphones provides a quick, easily accessible, and real-time payment method. One of the co-founders of M-KOPA, Nick Hughes, was one of the original creators the cellular service M-PESA (Nicke, 2013). Owned by telecom giant Vodafone, M-PESA is well established in Kenya, providing cellular service to 95% of Kenyan adults.

As shown through the interest of investors, the success of M-KOPA is promising so far. Access to electricity allows households to have light at night, which expands opportunities to work and study. People can also use units to charge cellphones and share with neighbors. Economic development researchers are also interested in the types of household data that could be collected. These opportunities are intriguing, and could potentially be useful ways that have yet to be thought of.

Jackson, Tom. "M-KOPA Solar raises $12.45m in latest funding round." Disrupt Africa. February 2, 2015.
http://disrupt-africa.com/2015/02/m-kopa-solar-raises-12-45m-latest-funding-round/
Mutegi, Lillian. Kenya: M-Kopa Wins Zayed Future Energy Prize. All Africa. January 15, 2015.
http://allafrica.com/stories/201501200298.html
Nique, Michael. A Look at M-KOPA: An interview with Nick Hughes. GSMA. February 14, 2013.
http://www.gsma.com/mobilefordevelopment/a-look-at-m-kopa-an-interview-with-nick-hughes
M-KOPA.com 1
M-KOPA Press Release. M-KOPA Solar Closes Fourth Funding Round. February 2015.
http://www.m-kopa.com/press-release/m-kopa-solar-closes-fourth-funding-round/

Steps in Solar: The Hydrogen Super Emitters
by Alison Kibe

For cheaper and more efficient solar cell technology, look no further than Picasolar Inc. Working with institutions like the University of Arkansas and the Georgia Institute of Technology, Picasolar is looking to make efficiency gains they've made using the hydrogen super emitter (HSE) process in the lab viable for commercial production levels.

A solar photo voltaic cell's most basic parts are a silicon wafer semiconductor, an anti-reflective coating, and font and back conductors (contacts) that complete a circuit with an external load. Picasolar recognizes that the system can be made more efficient by using a selective emitter structure where dopants are used to impose a lower resistance under the front contact—recognizable to most as the grid pattern on a solar cell—and a raised resistance in the areas between the front contact gridlines (Picasolar, 2013). The high resistance prevents recombination, which is important to the life of solar cells, and the low resistance improves conversion efficiency.

The problem Picasolar sees with selective emitters is that production requires extra steps to ensure precision during manufacturing (Picasolar, 2013). In addition, extra gridlines that block out more light must be used and are not cheap considering silver is the conductor of choice. HSE promises to overcome this by placing gridlines over a homogenous emitter to act as a mask while the areas between the gridlines are exposed to hydrogen. The hydrogen deactivates dopants between the gridlines that had created low resistance. This process maintains efficiency gains, eliminates production steps, and could reduce the use of silver by up to 20%.

"If successful, this approach represents the single largest technology leap in solar since 1974," Picasolar CEO Douglas Hutchings told University of Arkansas News in 2013. Since then Picasolar has raised $1.2 million and with a patent pending on its idea, is working toward putting the HSE processes into use (Branam, 2014).

Picasolar, 2013. Picasolar Technical White Paper.

Branam, Chris. "University-Affiliated Business Applies for Patent for Solar-Cell Efficiency Technology." University of Arkansas News. February 2013. http://news.uark.edu/articles/20249/university-affiliated-business-applies-for-patent-for-solar-cell-efficiency-technology

Branam, Chris. Picasolar Raises $1.2 Million in Equity Investments. University of Arkansas News. November 2014. http://news.uark.edu/articles/25992/picasolar-raises-1-2-million-in-equity-investments

One Step Closer to Cheaper Solar Power
by Briton Lee

Recently, researchers were able to produce organic photovoltaic (OPV) cells in a way that can be scalable to an industrial level. One of the barriers facing the widespread adoption of solar power is the cost-prohibitive nature of its production. Additionally, the conditions used to create inorganic solar panels, such as crystalline silicon are harsh; for instance, they must be produced at very energy-intensive high temperatures. Organic solar cells are being explored precisely because the organic materials characteristic of the product have a low production cost, with the added benefit of being flexible. Some of the drawbacks of organic solar cells are that they are not as durable as inorganic solar cells, and have a lower conversion efficiency. These drawbacks are attenuated by the potential of both scaling up the efficiency of the cells and robust mass production of organic solar cells with minimal resource input.

While it may have a relatively low conversion efficiency, we have to take into account the amount of energy it can potentially create relative to the energy input; the net return on energy is substantial. The Andersen group is the first to demonstrate the successful production of roll-to-roll, contiguous solar cells and confirms the possibility of producing a high technical yield with relatively simple conditions. The way these cells are produced involves the "printing" of multiple layers (14 in this case) one over another extremely quickly. Each 14 layer section can be printed in 1 sec, ultimately achieving a printing rate of 1.3 meters per minute. It is interesting to note that the majority of the research in OPVs focuses on high conversion efficiencies in small photovoltaic regions, neglecting the efforts in optimizing the average performance of large area devices, the latter of which are more relevant to practical implementation. The variety of applications for OPVs, such as packaging, flexible screens, clothing, make this material extremely viable in industrial settings, and these findings take us one step closer to the reality of widespread solar energy.

Andersen, T. R. et al. 2014. Scalable, ambient atmosphere roll-to-roll manufacture of encapsulated large area, flexible organic tandem solar cell modules. Energy Environ. Sci. 7, 2925–2933 (2014).

ScienceDaily (http://www.sciencedaily.com/releases/2009/04/090409151444.htm)

Royal Society of Chemistry (http://www.rsc.org/chemistryworld/2014/06/roll-roll-flexible-organic-tandem-solar-cells)

Measuring Impacts of Solar Development on Mojave Desert Plants

by Emil Morhardt

The massive development of wind and solar generating facilities in California's Mojave Desert puts California way out in front of the rest of the US in generation of renewable electricity, but at the same time the development drastically alters the desert ecosystem. Installation of photovoltaic arrays seems to require grading the land flat, removing all existing vegetation, and since there will be nothing to eat, all of the animals as well. To those who haven't travelled this wild desert during a verdant spring—something that happens only every few years—it might seem barren. But I've camped out in the middle of it many times in the spring when it is lush, covered with desert flowers, and alive with birds and other animals; to me it is the epitome of virgin wilderness. (My wife and I even wrote a book about it and took a lot of plant pictures...see reference below.) So, one question to ask is whether or not any of that desert life will recover under the solar panels.

A research project addressing a small part of this question has been under way in the Mojave since 2011, and while not ready for scientific publication, is turning up some interesting results. Tanner *et al.* (2014) constructed isolated solar panel simulators—angled sheets that shaded the ground and deflected rain—and have been checking to see if two of the tiny daisies that are often abundant after rains thrive under the panels. In 2011 when they started the experiment there were 850 Barstow woolly sunflowers (*Eriophyllum mohavense*) and 454 Wallace's woolly daisies (*Eriophyllum wallacei*) per square meter!

In 2012 there were almost none. (Rainfall makes all the difference in this ecosystem.) Last year there was enough rain to get some germination, so some early results are now available; the panels block out 85% of the solar radiation under them, and it is 11 °C cooler, so the researchers certainly expected differences under and away from the panels. They now know that the panels not only reduced emergence of the Wallace's daisies, more of them died before flowering. Also, there weren't as many other plant species under the panels, and the total density of plants was lower. They don't say anything about how much soil moisture there was under the panels compared to their control areas, but since this is probably the most important variable I'm sure they're monitoring it. Also, as any roadside botanist knows, it's the runoff from the edges of the road that often produce the only flowers, so comparing the runoff zone from the panels with the area under them should be interesting.

These are early results, but it is clear that the extreme year-to-year climatic variation in this environment is going to make it difficult to get definitive data, and it is going to take a long time. Meanwhile, many of the proposed solar panel facilities have already been constructed, so it will be possible to do similar research under the real thing. Will the operators let the desert vegetation grow under the panels? A drive through some of these facilities in the Antelope Valley the other day suggests not always, and the only vegetation in evidence where it was allowed was the aggressively invasive tumbleweed--not too surprising since all of the native vegetation had been graded away and the ground heavily disturbed. I hope that research on how to protect the desert biota under the panels becomes a standard condition of licensing.

Tanner, K., Moore, K, and Pavlik, B. 2014. Measuring Impacts of Solar Development on Desert Plants. Fremontia 42: 15-16

Morhardt, Sia, and Morhardt, Emil. 2004. California Desert Flowers; An Introduction to Families, Genera, and Species. University of California Press. 284 pages.

Solar Panels Might Not Help CO_2 Reduction Any Time Soon
by Emil Morhardt

The main considerations in whether and where to install photovoltaic (PV) panels are how much sun there is, and how much the panels cost. Right? Not necessarily. Engineers at Arizona State University have just published a paper pointing out that if a goal of installing photovoltaics is to decrease greenhouse gas emissions, it would be prudent to consider the emissions from manufacturing—which vary significantly by panel type—how long they stay in the atmosphere, and whether or not the installation is competing with other renewable energy sources rather than with fossil fuel burning. Because of the greenhouse gases associated with manufacturing, all panel installations increase greenhouse effects in the short term, although the initial two-year effect is to reduce them owing to sulfur and nitrogen oxides released from power plants during manufacture.

In their lifecycle modeling, Ravikumar *et al.* (2014) concluded that in California and Wyoming it takes at least six years to get to the point that the panels are reducing the greenhouse effect, and it might take as long as eleven years in Wyoming and twelve years in California. This five-to-six-year difference has to do with the choice or technology and the primary energy source where the panels are installed. Chinese polycrystalline panels use more energy to refine the panel components than do the cadmium telluride panels made in Malaysia; China gets most of its energy from coal, whereas

Malaysia gets about half of its energy from the less-polluting natural gas, and 10% of it from hydroelectricity; and installations in California are competing with the equally-renewable hydroelectric energy which makes up a significant part of the electricity on the grid; Wyoming is more dependent on coal. In all cases however, the model calculates that the sooner photovoltaics are deployed, the better.

This blog summary greatly simplifies a complex and insightful lifecycle modeling effort that seems to me to serve as a good introduction to the process for the uninitiated. I'd recommend reading the whole paper.

Ravikumar, D.T., Seager, T., Chester, M., Fraser, M.P., 2014. Intertemporal Cumulative Radiative Forcing Effects of Photovoltaic Deployments. Environmental Science & Technology dx.doi.org/10.1021/es502542a Abstract and a figure at: http://bit.ly/1AF5fGK

One of the Nation's Biggest Solar Farm Opens in California
by Shannon O'Neill

The Desert Sunlight Solar Farm located in Riverside County, opened in February 2015 as one of the biggest solar farms in the world. First Solar, who also contributed more than 8 million solar modules to the project, runs the project. The farm has 4,000 acres of solar panels, providing the capability to produce 550 MW. This is enough to provide energy to more than 160,000 homes. Additionally, this energy source will replace the use of 300,000 tons of CO_2 each year, a number equivalent to removing 60,000 cars off of the roads. In addition to the environmental benefits, the project has also created many jobs. This project is aiming to contribute to governor Jerry Brown's initiative of one-third of California's energy coming from renewable resources by 2020, and one-half from renewable sources by 2030.

This project opens during a time where the future of solar energy is uncertain due to the fact that federal funding and investors' interests have decreased in recent years. Specifically, the federal investment tax credit is expected to decrease from 30% to 10% by the end 2016. Additionally, with many states already on track to meet renewable energy goals, investing in solar energy has not been a priority. However, as solar energy is slowly becoming price-competitive due to the decrease in prices of photovoltaic panels along with the opening of this solar farm, there is hope of re-initiating such interests in solar energy.

Pacific Gas and Electric Company and Southern California Edison have already agreed to purchase energy from the Desert Sunlight Solar Farm for the next twenty years. Additionally the Obama Administration is making renewable energy a priority. They have designated 22 million acres in California for the sole use of renewable development in order to generate 20,000 megawatts of power by 2020. This is enough energy to power around 6 million homes.

Huge Solar Farm Opens in California: Enough Energy from 160,000 Homes (http://www.latimes.com/local/lanow/la-me-ln-solar-farm-20150209-story.html)
Desert Sunlight Solar Farm (http://www.firstsolar.com/en/about-us/projects/desert-sunlight-solar-farm)
550 MW Desert Sunlight Solar Farm in California Now Online (http://cleantechnica.com/2015/02/10/550-mw-desert-sunlight-solar-farm-california-now-online/)

Debate over Buying or Leasing Solar Panels in Residential Homes
by Shannon O'Neill

In recent years, the focus on the transition to the use of more renewable energy along with the fact that solar panels are becoming more price-competitive has led to their greater prominence in residential homes. Specifically, more than 600,000 homes have solar panels today. This has lead to many homeowners to question whether it is more cost effective to buy or lease solar panels.

NPR took a look at two homeowners from Maplewood, New Jersey. The two are neighbors with similar homes in size and style, however, one family decided to lease solar panels while the other decided to buy. Their choices highlight the drawbacks and benefits from each decision, and give evidence as to why some options are better for some than others.

The allure to leasing solar panels is the lack of up-front cost. Roebuck, the homeowner who decided to sign a 20-year lease agreement for the solar panels, pays $69.25 per month to a company that installed and maintains the panels. He explained that this cost replaces his monthly electricity bill. With that, he explained that he was quite certain his electricity bill would have gone up beyond this expense had he not leased the solar panels. However, because he has leased the solar panels, he does not receive the government subsidies from switching to renewable energy. Rather, the company that he leases from does.

Roebuck's neighbor, Ebinger, bought a solar panel system for her home for a cost of $35,000. With the purchase she received a 30% federal tax credit, which reduced the initial price, along with monthly savings on her electricity bill. Additionally, once she accumulates one thousand kilowatt-hours of energy, she can sell one Solar Renewable Energy Certificate for a fluctuating price around two hundred dollars. Ebinger stated that she sells around seven of these certificates a year. Following this current trend, Ebinger should have free electricity for the next fifteen years. However, this requires Ebinger to have a detailed spreadsheet that analyzes the expenses and payback of the solar panels, and therefore requires a lot more work.

Roebuck believed he made the right decision for himself, as leasing allowed a low-risk option to renewable energy, whereas Ebinger enjoyed having more control of the costs and benefits throughout the ownership of the solar panel.

John Farrell works with the Institute for Local Self-Reliance, an organization that has looked heavily at the debate between buying and leasing solar panels. Their website includes a calculator to help homeowners chose the best option for them. The overall findings show that a consumer saves an extra $2,200 over the panels' thirty-year life span by purchasing them. This leaves Farrell to encourage homeowners to buy solar panels if they can afford it, but leasing is still a good alternative.

NPR: The Great Solar Panel Debate: To Lease Or To Buy?
(http://www.npr.org/2015/02/10/384958332/the-great-solar-panel-debate-to-lease-or-to-buy)

Growing Competitiveness of Solar Energy
by Shannon O'Neill

Until recently, the hefty costs associated with solar panels have made usage heavily reliant on tax breaks and subsidies to reduce the cost. This was made possible by projects like the California's Solar Initiative and the federal Investment Tax Credit. However, the cost of solar panels has dropped dramatically—from $150/watt in 1970 to 60 cents/watt today. Additionally, the installation cost has decreased from $10/watt of generating capacity to just $5/watt, today. This has allowed solar energy to become less reliant on such projects, which subsequently has lead its ability to out-compete conventional energy sources.

The growing price-competitiveness of solar energy has lead to Californians installing close to 2,000 megawatts of solar panels since 2007. Additionally, solar energy was able to create jobs throughout the recent recession. For example, OCR Roofing, located in Sacramento, California, trained its employees to install solar panels on rooftops, in lieu of laying-off workers and in order to continue business throughout the recession. This allowed the company to become one of the "largest privately owner solar and roofing companies in the nation" (CNBC). In California alone, the solar energy sector is responsible for employing over 50,000 people.

Additionally, companies like SolarCity and SunRun have been able to decrease the upfront cost of installing solar panels for homeowners. They allow the systems to be leased, which allows homeowners' to pay a monthly price that is typically lower than their normal utility rate, making the purchase worthwhile.

Some SolarCity customers have gone on to add additionally battery backup systems that will eventually allow homes to be powered through blackouts and to eventually drop off of the energy grid all together. Such batteries will also reduce the issues associated undersupply of renewable energy during peak times of day as they can store energy and dispatch it to the grid when it is needed.

Finally, the drought in Western United States has reduced the amount of kilowatts produced by the hydroelectric dams and reservoirs in the area. This has led to growing electricity rates, further displaying the growing demand for solar energy.

CNBC: "The big energy debate that solar power has finally won"
http://www.cnbc.com/id/102602378

Community Solar Gardens and Other Shared Solar Power Systems Expected to Rise in 2015
by Melanie Paty

Community solar gardens were originally popularized in Colorado and now have spread across the nation to several states. A solar garden is a site of panels that connects to the local power grid with the community sharing ownership. Ten states currently promote ways to share renewable energy and even more are promoting community solar. California passed new legislation in late January mandating three of the largest utilities to contract for 600mW of new solar capacity. This June, Clean Energy Collective, a Colorado based community clean energy developer, will be developing three community solar projects in Massachusetts that are expected to power 400–500 residential and commercial units. In 2014, Minnesota passed a community solar garden law, which was designed to encourage small-scale developments, but most the proposals have been for utility scale projects made up of many neighboring 1mW plots. A local energy company is concerned that the grid cannot handle such a large increase in electricity.

Community solar opens up the clean market to new customers who could not otherwise participate due to cost or other logistical issues such as home renters who do not own property and therefore cannot install panels. Furthermore, rooftop solar panels are only compatible with one fourth of United States residences, whereas solar gardens are open to all. Customers can subscribe to the power of the community solar garden and receive credit on their utility bills. Some estimate that solar energy bills will drop below normal rates in the next few years, while others estimate that savings will start immediately. The CEO of SunShare, a Colorado-based solar garden developer, predicts that solar gardens will become mainstream in 2015.

Karnowski, Steve. ABC News "New Concept in Solar Energy Poised to Catch on Across US" March 7th 2015. (http://abcnews.go.com/US/wireStory/concept-solar-energy-poised-catch-us-29463892?singlePage=true)

The Cutting Edge of Solar
by Chad Redman

Solar panels are steadily gaining traction as a common source of renewable energy. However, keeping track of the latest and greatest technologies in this field can be difficult given the high level of research dedicated to developing cheaper, more efficient solar cells. A great diversity of solar technologies currently exists, though a few in particular stand out as the most viable options for residential and commercial use.

Spectrolab, a Boeing subsidiary, presents perhaps the most advanced solar panel technology manufactured today – multijunction concentrator photovoltaic panels. Among a cluttered and competitive industry, Spectrolab's multijunction CPVs stand out as the most efficient solar cells available today.

CPVs operate by using mirrors and other optics to capture sunlight and focus it onto a solar panel, typically tracking the sun throughout the day and requiring cooling systems to ensure reliability of the solar cell. Multijunction photovoltaic cells describe cells that use multiple semiconductors, essentially allowing the cell to convert a broader range of light wavelengths into electricity. This allows multijunction cells to convert light energy into electricity at a higher efficiency as compared to single semiconductor photovoltaics.

Spectrolab primarily produces cells for applications in space, meaning that its products must be extremely reliable and effective, if not particularly cheap. What makes Spectrolab's multijunction CPVs so remarkable, however, is the level of efficiency they have achieved. Typical PV cells used in domestic applications achieve efficiencies of less than 20%. Concentrating photovoltaics from Spectrolab have achieved a record 41% efficiency. This is the most efficient technology currently available, and will lead the way to more efficient technology for household use.

"New Frontiers in Solar Cell Conversion Efficiency." Spectrolab. N.p., n.d. Web. March 23. 2015.

A Spherical Solution to Solar Power
by Chad Redman

Solar energy is one of today's most promising sources of renewable energy. A mountain of research and development has focused itself on developing effective technology for capturing this abundant, if diffuse and intermittent, source of energy. As a result, the solar industry now boasts a huge array of solar cell and solar thermal products, with significant political support to back its proliferation. However, the efficacy of solar power has always suffered from low efficiency, intermittency, and high cost.

Fortunately, one innovative company called Rawlemon is tackling these challenging flaws in solar energy collection. Gizmag writer Stu Robarts paints a confidence inspiring picture of Rawlemon and its products, beginning with founder and inventor Andre Broessel. Allegedly, Broessel was inspired in his design by his daughter's toy marbles to create a spherical concentrator lens which could focus sunlight onto a small photovoltaic cell. The end result of this inspiration is a beautiful and effective transparent lens, filled with water, which can capture diffuse sunlight and focus it onto a relatively small PV surface.

In particular, Rawlemon offers two models of their technology which implement the water-filled lens orbs. These models achieve a level of efficiency far greater than that of stand-alone PV cells due to their ability to capture both solar and thermal energy. Thermal energy is collected in the water contained within the lens. Additionally, the Rawlemon models boost their efficiency through the use of multi-axis tracking bases, allowing optimal energy collection all day long.

The Rawlemon solar technology is particularly exciting due to its high efficiency and ability to store thermal energy in the lens. With innovation like this, alternative energy sources seem ever more capable of replacing fossil fuels.

http://www.gizmag.com/rawlemon-spherical-solar-energy-generators/30453/

The Fight for Home Solar Systems
by Abigail Wang

When people say they want to "get off the grid" it usually means getting off social networks. Hawaiians, however, are giving it a whole other meaning; they want to get off power distribution grid. Several residents are finally getting their personal solar system applications approved.

In a solar-rich area like Hawaii, more people are eager to make their own electricity with rooftop solar panels rather than pay expensive monthly bills. In fact, rooftop systems are on about 12% of Hawaii's homes, which is the highest proportion in the United States.

After a long debate since 2013, state energy officials ordered one of the biggest power companies on the island, Hawaiian Electric Company, to approve a long backlog list of solar applications. The company has barred customers from getting personal solar systems, claiming that the power generated by rooftop systems was too much for it to handle on the grid.

Utility companies are afraid of losing money as more people implement solar panels in their homes. Edison Electric Institute, an association that represents electric companies, warned its members that the demand for electricity is decreasing while home solar systems are rapidly spreading. In response, companies have reduced incentives to green energy alternatives, added additional fees to utility bills, or pushed home solar companies out of the market. For instance, in February, Salt River Project, a large utility company in Arizona, began adding $50 to a typical monthly bill for new solar customers. In Wisconsin, regulators approved fees that would add $182 a year for the average solar customer.

Solar companies aren't going down without a fight. Several are using regulators, lawmakers, and courts to question utility companies' authority and promote solar energy. Installers for rooftop solar panels advise customers to get off the grid entirely by putting in batteries along with solar panels so that they can store power generated during the day to use at night. This alternative is more expensive, but the customer doesn't have to rely on the utility's network of power lines.

The battle is far from over between power companies and the public. As more people opt for home solar systems, utility firms will have to adjust their businesses to adapt to these homeowners.

Cardwell, Diane. "Solar Power Battle Puts Hawaii at Forefront of Worldwide Changes." New York Times: 19 April 2015.
http://www.nytimes.com/2015/04/19/business/energy-environment/solar-power-battle-puts-hawaii-at-forefront-of-worldwide-changes.html?_r=0
Lippert, John and Christopher Martin. "Musk's Cousins Battle Utilities to Make Solar Rooftops Cheap." Bloomberg Business: 14 April 2015.
http://www.bloomberg.com/news/articles/2015-04-15/elon-musk-s-cousins-battle-utilities-to-make-solar-rooftops-cheap

SOLAR THERMAL

Another major use of solar power part from photovoltaics is to convert the energy in sunlight directly into heat, and use that to create high-pressure steam to drive the steam turbines typical of a gas- or coal-fired power plant. The most impressive new applications (though working prototypes have been around for decades) are power towers of the type shown on the cover of this book: mirrors concentrate the light on a central boiler t the top of the tower and either create steam directly, or heat up a working fluid that is used to create steam later. For a while it looked like most of the new solar power installations in California's Mojave Desert would be of this sort, the most impressive of which is the set of three towers in Ivanpah Valley, discussed blow. The radical decrease in cost of photovoltaic panels, though, may be short-circuiting the development of this type of solar power.

BrightSource's Ivanpah CPS Bird Fatalities Controversy
by Mariah Valerie Barber

In February 2014, nearly a year ago, BrightSource Energy, Inc.'s Ivanpah concentrating social power (CSP) plant officially opened in Southern California's Mojave Desert, after four years of construction. The Ivanpah power plant uses heliostat, software controlled mirror technology to concentrate the sun's solar rays and direct them to a water tower. The concentrated sunlight is then reflected onto boilers that create steam which is used to generate power by utilizing a turbine. Ivanpah is currently the world's largest concentrated solar plant, occupying around five square miles, with 173,500 large "garage-door" sized heliostat mirrors directed to central power towers. BrightSource developed Ivanpah and is now partially owned by Google and NRG Energy. Currently Ivanpah is working towards reaching its full energy producing capacity. Once Ivanpah is operating at full capacity it will be able to generate 140,000 homes annually.

Despite Ivanpah being recognized internationally for its historically step towards clean energy and its receipt of POWER's 2014 Plant of the Year Award, BrightSource has been receiving a great deal of criticism because of the negative impact it has had on the Mojave Desert wildlife (powermag.com). In building Ivanpah, BrightSource, NRG, and Google invested $25 million dollars into mechanisms to prevent the desert tortoise that live in the area surrounding Ivanpah. However, the major problem has been with birds flying into the solar rays or crashing into the heliostat mirror panels. Some sources such as the Center for Biological Diversity estimate that Ivanpah will lead to the death of 28,000 birds a year. According to BrightSource's website, however, only 321 bird fatalities were recorded between January 2014 and June 2014 which would put their annual bird fatality estimations at 1,000 deaths per year. The area surrounding the plant can get as hot as 1000-degrees Fahrenheit, so surrounding birds are being burned to death when flying through the plant. The resulting death of the birds near Ivanpah is being referred to as a result of "solar flux". Due to such criticisms of the Media, Ivanpah has responded with "The Top Five Things Some Media Can't Seem to Remember About Ivanpah" on BrightSource's official blog, specifically addressing the bird fatalities as, "Ivanpah's impact on Birds is Minimal, But Also a Priority." BrightSource's technology will provide the state of California with a great deal of clean energy, but can it do so while minimizing its harm to local birds?

BrightSource Energy, Inc. (http://www.brightsourceenergy.com/)
http://www.powermag.com/ivanpah-solar-electric-generating-system-earns-powers-highest-honor/
http://www.brightsourceenergy.com/the-top-five-things#.VMgGfXDF8lS
http://breakingenergy.com/2014/08/27/a-solar-bird-death-story-ignites-controversy/#comments
http://www.brightsourceenergy.com/how-it-works#.VMgGAHDF8lQ.

Crescent Dunes Solves the Solar Farm Bird Death Problem with Energy Storage
by Liza Farr

Large-scale solar farms have been built in the past few years in order to generate a large amount of energy from solar power in one plant. These farms entail large fields of mirrors that are computer-controlled to reflect light to the top of a tall tower (Danko, Feb 10, 2015). By using these mirrors, solar power is concentrated at the top of this tower, where various methods have been used to harness the light into thermal energy (Danko, Feb 10, 2015). The most well known project to date is Bright Source in Ivanpah Valley, California, which was backed by $1.6 billion in federal loan guarantees (Danko, Feb 10, 2015). In this plant, the concentrated solar light directly boils water, but there is no method for energy storage. Mirrors must go into standby mode during the day when there is too much power supply compared to the use (Kraemer Apr 15, 2015). These mirrors create solar flux, which is concentrated sunlight radiated into a certain area (Kraemer, Apr 15, 2015). When these mirrors go into standby, they are shifted from focusing on the top of the tower to focusing on a concentrated area in space, and when birds fly through this area, many die from the heat (Kraemer, Apr 15, 2015). In the first six months of operation at Ivanpah, 321 birds and bats were killed (Kraemer, Apr 15, 2015).

SolarReserve is another company that has received $737 million in federal loan guarantees to build a solar farm called Crescent Dunes (Danko, Feb 10, 2015). Crescent Dunes originally faced the same problems with bird death as Ivanpah, but a simple fix has allowed this project to reduce bird deaths to zero in three months (Kraemer, Apr 15, 2015). The engineers on the project were the first to discover the solar flux that occurred when multiple mirrors went into standby mode was what killed the animals (Kraemer, Apr 15, 2015). They noticed then that during testing, these mirrors will often sit in standby mode for several hours, but during normal operation, the mirrors at Crescent Dunes will not need to enter standby mode as Ivanpah mirrors did. The main innovation of the project is energy storage, which allows the mirrors to remain focused on the receiver all day since excess heat is simply stored. This also means the plant will not need to supplement with natural gas during periods of low sunlight, as Ivanpah must (Danko, Feb 10, 2015).

The main energy storage innovation of Crescent Dunes comes from their method of capturing the solar thermal energy. Rather than boiling water directly, the concentrated sunlight heats a mixture of sodium and potassium nitrate that is used to either super heat water, or be stored in insulated tanks to use during cloudy weather and nighttime (Danko, Feb. 10, 2015). The temperature of the salt is 300 degrees higher than another solar farm using molten salt, Solana generating station in Arizona, because they also don't use any kind of transfer fluid to heat the salt—they heat it directly with the sunlight (Danko, Feb 10, 2015). Molten salt thermal power was originally developed by rocket scientists over 20 years ago, and SolarReseve has commercialized the technology (Santa Monica Observer, Apr 13, 2015). This technology allows the power tower to operate 24/7, giving it the stability and predictability of traditional fossil fuel energy sources (Santa Monica Observer, Apr 13, 2015). The salt never needs replacement during the over 30-year lifetime of the power plant, although there are still concerns about battery performance, degradation, and end of life environmental impact (Santa Monica Observer, Apr 13, 2015). Crescent Dunes will be able to store up to 1,100 MWh, equivalent to 10 hours of full output, and power more than 75,000 homes through generating 500,000 MWh per year (Danko, Feb 19, 2015 and Santa Monica Observer, Apr 13, 2015).

Unfortunately, many tax credits for new construction of renewable energy are due to run out, slowing the growth of solar farms and other renewable energy plants. President George Bush's 2008 stimulus package included an eight-year extension of the 30% investment tax credit for qualifying renewable energy projects (Danko, Dec 23, 2014). In 2009, President Obama allowed project developers to forego the somewhat cumbersome investment tax credit for a lump sum payment once a project is up and running, but the payments expired at the end of 2011 (Danko, Dec 23, 2014). Projects started before then will still receive the money, but if future renewable energy companies will be able to build projects such as the Crescent Dunes farms, which have made great advances in efficiency and capacity, more help with capital cost must be provided by the federal government.

Kraemer, Susan. Solar flux solution brightens future of concentrated solar power. April 15, 2015.
[http://www.renewableenergyworld.com/rea/news/article/2015/04/solar-flux-solution-brightens-future-of-concentrated-solar-power]

Danko, Pete. SolarReserve: Crescent Dunes Solar Tower Will Power Up in March – Without Ivanpah's Woes. February 10, 2015.
[http://breakingenergy.com/2015/02/10/solarreserve-crescent-dunes-solar-tower-will-power-up-in-march-without-ivanpahs-woes/].

Danko, Pete. 1603: A big renewable energy subsidy, Yes; Just Don't Call It a Bailout. December 23, 2014. [http://breakingenergy.com/2014/12/23/1603-a-big-renewable-energy-subsidy-yes-just-dont-call-it-a-bailout/].

Santa Monica Observer. SolarReserve Changes The Industry. April 13, 2015.
[http://www.smobserved.com/story/2015/04/13/business/solarreserve-changes-the-industry/1083.html].

Solar Plant Development Impeded by Tax Credit Reductions

by Niti Nagar

Solana, a solar thermal power plant in the Arizona desert has been functioning for more than a year now. Developed by the Spanish energy and technology company Abengoa, it has succeeded in producing energy when the sun is not shining, allowing solar production to continue at full capacity long after 6 o'clock at night. According to Brad Albert, the general manager for resource management at Arizona Public Service, this feature adds a lot of value because the customer demand for electricity is so high after the sun goes down. The Solana uses parabolic mirrors that direct sunlight on pipes that carry the heat to tanks of salt. The heat is contained for up to six hours until the plant converts it to make electricity via steam.

Abengoa opened a similar plant on January 23, 2015 in the Mojave Desert in California, which is expected to supply enough power for 91,000 homes. However, despite the success of the technology, Abengoa and other power plant developers are hesitant to build more solar plants in the United States. This is due to the large uncertainty regarding an Investment Tax Credit worth a whopping 30% of the project's cost. The subsidy is set to remain until 2016, but this does not give developers enough time to secure land, permits, financing and power-purchase agreements. The Solar Energy Industries Association hopes to lobby Congress to extend the credit beyond 2016 before it drops to 10% in 2017. Despite the technology feasibility to offer an alternative form of energy in the United States, government policies and the lack of tax incentives are forcing developers such as BrightSource to focus on other countries, including China and Israel.

However analysts say some solar companies have responded well by implementing cost reductions in their technology development. To further spur interest from investors, advocates are urging the Obama administration to extend master limited partnerships to include solar projects in real estate investment. By expanding solar energy to different markets, companies can take advantage of tax benefits that would otherwise be unavailable. According to former Senate Finance Committee staff member, Paul Bledsoe, Congress is keen on eliminating many of the 42 existing energy tax subsidies with Republicans largely against subsidies for renewables and Democrats largely against those for fossil fuels. Now that the technology exits, the overarching question is whether the government can find a way to effectively phase out incentives over the years (perhaps will a gradual reduction rather than a steep 20% decrease) and replace them with other technology-neutral approaches.

Cardwell, Diane. "Worry for Solar Projects After End of Tax Credits." The New York Times. The New York Times, 25 Jan. 2015.
http://www.nytimes.com/2015/01/26/business/worry-for-solar-projects-after-end-of-tax-credits.html

HYDRO/TIDAL/WAVE ENERGY

Hydroelectric power, a logical outgrowth of water wheels, is well developed throughout much of the world and it is often noted that most of the feasible large hydro projects have already been built. Not so for wave and tidal power. Just offshore these are vast sources of energy waiting to be tapped, and for all intents and purposes, are completely unexploited. Why? Because building machines that will withstand the marine environment is difficult, what with its corrosive waters and sprays and occasionally violent demeanor, capable, as we have witnessed recently in Fukushima, of disrupting even heavily engineered and massively armored nuclear power plants. But things are progressing, and the rewards will be great once a few stable technologies evolve. For now, though, it is early days.

Clean Current Utilizes Marine Tidal Turbines to Produce Renewable Energy
by *Mariah Valerie Barber*

Clean Current Power Systems Incorporated is a private company based in British Columbia that focuses on hydrokinetic power generation. Specializing in marine energy engineering, Clean Current was the first company to use a tidal turbine or marine turbine energy. Clean Current's tidal turbine utilizes the same basic framework used by the standard river in-stream turbine, only the company has incorporated bi-directional technology that allows the turbine to change directions automatically depending on the movement and direction of the tides. Clean Current's tidal turbines are predicted to last up to 25 years and are able to be placed in marine areas 7–25 meters deep.

The company's first tidal turbine was used at the Race Rocks Environmental Reserve off of Vancouver Island, Canada. Race Rocks, a nature reserve needed a self-sustainable green energy source so the tidal turbine was installed so that research could be done from the reserve. The first tidal turbine was installed in May 2006 and then removed five years later so that it could be placed in Canadian Science and Technology Museum in Ottawa, Ontario, Canada in March 2011. The first installation was considered a success. Currently Fundy Tidal Incorporated, another hydrokinetic energy company and Clean Current Power Systems are working together to install and test another tidal turbine in the Grand Passage, Nova Scotia. The new tidal turbine will be part of an entire Tidal Power System, which also allows for the storage of energy. The new installment of another tidal turbine is expected to last a total of 12 months and begin during the spring of 2015. Although the official date has not yet been released, the project, if everything goes according to schedule, should start in April 2015.

Clean Current, Inc. (http://www.cleancurrent.com/)

Tidal Lagoon Power Reveals Plans to Build World's First Lagoon Power Plant
by Nour Bundogji

On Match 2, 2015, Tidal Lagoon Power made a significant step towards the delivery of full-scale tidal lagoon infrastructure in the UK—building the world's first lagoon power plant. Tidal Lagoon Cardiff will consist of six giant structures—four will be built in Wales, and two in England—which will harness powerful coastal tides and generate as much as 8% of the UK's total power. For an investment of £30 billion, these lagoon power plants could also power all Welsh homes.

How does tidal lagoon power work? BBC environmental analyst Roger Harrabin explains, "The lagoons operate a system similar to a lock gate to alter the water level on either side of a sea wall. When the tide is full outside the lagoon, the gates are opened and water rushes past the turbines to fill up the lagoon. When the tide turns to go out, the gates are shut to hold the water inside the lagoon. As low tide is reached outside the wall, the gates are opened to generate power again as water flows through from the raised water level in the lagoon."

The UK government and power firms are keen on backing this renewable energy project because, unlike power from the sun and wind, lagoon-energy is predictable.

Although there is reservation that fish could be sucked into turbines, Tidal Lagoon Power assures that the lagoons will ultimately benefit local ecosystems by serving as artificial reefs.

Another issue that will be resolved soon is how much the company plans to charge the government for the electricity the lagoon generates. Currently, the Hinkley nuclear station charges £92.50 per MWhfor its power. Thus, Harriban suspects that Tidal Lagoon Power will charge somewhere between £90 and £95 per MWh for its plant.

A planning application for the project is expected in 2017. If approved, it could be generating power by 2022 giving the government a sustainable source of clean energy that has little to no risks associated with it.

The cost would be funded by electricity bill-payers under the existing government scheme to promote home-grown, low-carbon energy.

Mr. Davey told BBC News: "I'm very excited by the prospect of tidal power. We have got some of the biggest tidal ranges in the world and it would be really useful if we could harness some of that clean energy."

Brian, M. UK to build the world's first tidal lagoon power plants. Engadget.com. March 2, 2015.
http://www.engadget.com/2015/03/02/worlds-first-tidal-power-plants-uk/
Tidal Lagoon Cardiff. Tidal Lagoon Cardiff could power all Welsh homes for 120 years. March 2, 2015.
http://www.tidallagoonpower.com/h/news/starting-pistol-fired-for-delivery-of-fullscale-tidal-lagoon-infrastructure/bp127/
Harrabin, R. World's first lagoon power plants unveiled in UK. BBC.com. March 2, 2015
http://www.bbc.com/news/science-environment-31682529

Carnegie Wave Energy
by Dylan Goodman

Carnegie Wave Energy is an Australian-based energy production company. Over the past 15 years they've been working to develop their patented CETO technology, a technology that allows for the conversion of wave energy into electrical energy. The project initially began in 1999 and seven years later they released their first prototype. Having raised over $80 million in funding, the project has been in the developmental stages until just recently. In February 2015 the first on-grid wave-powered energy plant was activated off the coast of Western Australia. While Carnegie Wave Energy has been operating prototypes since their initial 2006 release, "This is the first array of wave power generators to be connected to an electricity grid in Australia and worldwide." (Ivor Frischknecht – CEO of the Australian Renewable Energy Project) The Australian Renewable Energy Agency has provided significant funding for the completion of this project. CETO, the energy conversion technology, transforms swells from waves into zero-emission power as well as zero-emission desalinated water. Thus, two-part technology is capable of producing both electricity as well as fresh water. The CETO technology differs from other wave-energy technologies in that the majority of the process occurs underwater, making it more resistant to various weather conditions. The system operates through a series of pumps installed on the ocean floor. The pumps are tethered to floating buoys on the service, and waves crashing into the buoys drive the pumps by forcing seawater into a pipeline beneath the ocean floor. Like other forms of power plants, the pressure generated from the water is used to drive turbines and produce electricity. Alternatively, "The high-pressure water can also be used to supply a reverse osmosis desalination plant, replacing or reducing reliance on greenhouse gas-emitting, electrically-driven pumps usually required for such plants," according to their website. The project will supply both fresh water and power to HMAS Stirling, Australia's largest naval base. There are currently two CETO wave units which have been successfully operating for over 2,000 hours, while newer, larger units are being developed.

Gough, Myles. "World's First Grid-connected Wave Power Station Switched on in Australia." ScienceAlert. N.p., 23 Feb. 2015. Web.

Converting Low-Head Hydropower into Air Pressure/Electricity
by Emil Morhardt

Low-head hydro, in this case 2 meters, is not normally viewed as a good source of energy for electricity production, but a clever paper from the engineers at Lancaster University, in the UK, suggests using a shore-based siphon to generate air pressure. Although they didn't try converting the air pressure to electricity, they certainly could if the economics were right, and it looks as if they might be. As the researchers point out, there are other alternatives to water turbines in current use, including Archimedes screws, hydro-venturis, and water wheels, all of which minimize harm to fish, but this approach might be even cheaper. The idea is to siphon water from above a small dam (weir) to below it, and entrap air into the water stream through a tube at the top of the siphon. The air gets compressed in the process and can be bled off for whatever use is handy; generating electricity with it might not be the best use since it will involve energy losses that might be avoided if the air were used directly, say to run some kind of pneumatic machine.

The engineers managed to capture 63% of the energy of the change in water elevation, which seems pretty good to me. They also note that since there are no moving parts inside the siphon, it might make a good fish passage facility as well, something not true of any of the competing technologies.

Mardiani-Euers, E., 2014. An Alternative Approach in harvesting Low Head Hydropower using a Siphon System by converting Water Power into Air Pressure, Proceedings of the 3rd Applied Science for Technology Innovation, ASTECHNOVA 2014, Yogyakarta, Indonesia.
http://tf.ugm.ac.id/astechnova/proceeding/Vol3No1/Astechnova_2014_1_08.pdf

Will Windpower Increase Hydroelectric Environmental Impacts?
by Emil Morhardt

Hydroelectric projects can be terrific for meeting peak electricity load demands; if they store water in reservoirs, they can release it more-or-less instantly to generate electricity just in time to meet the demand. This is what pumped storage hydroelectric facilities are designed to do from the start, usually pumping water uphill from one reservoir to another to store energy, then letting it flow back down when energy is needed. The only potential significant environmental impact from this operational phase would result from reservoir water-level changes during the cycle. In a non-pumped-storage situation where the reservoir is behind a dam on a river and the peaking strategy with the best economics is to release practically no water until needed for peaking, then to release a lot, there are plenty of potential downstream environmental impacts. Such a strategy is utilized in the mid-Atlantic US by some utilities. In a paper just accepted by the journal Environmental Science & Technology, researchers at the University of North Carolina and Duke University looked at releases at Roanoke Rapids Dam on the Roanoke River and tried to figure out if adding wind to the mix of renewable power would increase or decrease these potential impacts (Kern *et al.* 2014). Despite earlier suggestions that it would, they decided not, based on the model results that predicted very little increase in the downstream "flashiness" over current operational conditions, even with 25% new wind market penetration.

Kern *et al.* used an electricity market (EM) model for the Dominion Zone of PJM Interconnection, a 23 GW electric system, to which the Roanoke Rapids Dam can contribute, at most, 100

MW. The system gets its energy from a mix of coal, natural gas combustion and combined cycle turbines, nuclear, pumped storage hydropower, conventional hydropower, biomass, and oil. Pumped storage hydropower provides 6.9% of the power, but hydropower of the sort they are interested in provides only 2.1% of it, and the dam in question could provide less than half a percent of the overall mix when operating at full capacity. Adding wind power of up to 25% average annual market penetration [a lot] could conceivably substantially increase the need for peaking hydropower to make up for unexpected intermittencies in the wind, leading to very high real-time electricity prices, which might then encourage a much flashier operation of the Roanoke River Dam to maximize revenues there. It turns out that if the wind market penetration were as high as 25%, there would be a modest increase in downstream flashiness, but the authors figure that the environmental impacts of building the dam in the first place, and those of its current operation, far outweigh those due to any additional effects of even 25% windpower penetration.

Kern, J., Patino-Echeverri, D., Characklis, G.W., 2014. The Impacts of Wind Power Integration on Sub-Daily Variation in River Flows Downstream of Hydroelectric Dams. Environmental Science & Technology. (DOI: 10.1021/es405437h, Accepted for publication) http://bit.ly/1knbCuD

Micro-Hydroelectricity in Building Water-Supply Pipes
by Emil Morhardt

Seems like it's getting to the point that no possible source of power should go unharvested. Chang *et al.* (2014) envision the water tanks at the top of apartment complexes in Taiwan as mini-pumped storage projects: by installing miniature turbines in the water supply pipes feeding the building from these tanks, electricity can be generated whenever the occupants use any water. The pipes are 4–6 inches in diameter, and a single turbine can generate about 3 watts under the expected water flows. The experimental turbine blades were printed on a 3-D printer until the engineers got the result they wanted; a set of three airfoil blades that didn't alter the flow rate of the water (*i.e.*, extracting this amount of energy apparently didn't interfere with the functioning of the water supply system). The authors figure that they could get enough electricity out of a building's water supply lines to run a few light fixtures. They didn't explore it much, but the drains are another obvious source of potential energy. This all seems good.

Chang, C.-Y., Huang, S.-R., Ma, Y.-H., Hsu, Y.-S., Liu, Y.-H., 2014. The Feasibility of Applying Micro-hydroelectric Power Technology in Building Water Supply Pipes. Scientific-Journal.com. http://www.scientific-journal.com/articles/environmental/3/11.pdf

Fishing Boat Transformed to Harness Energy from Ocean's Waves
by Niti Nagar

It might be possible to harness the movement associated with the ocean's natural waves using docked fishing vessels. Researchers are demonstrating that this simple idea is feasible using a demonstration vessel currently docked offshore in Western Norway.

Transforming a vessel into a wave power plant requires installing four large chambers, in the vessel's bow, each equipped with an air-powered turbine. As the waves strike the vessel, the water levels in the chambers rise leading to an increase in air pressure, which consequently drives the four turbines. The chambers respond to different wave heights, allowing greater wave heights to contribute more air pressure. Each turbine is capable of produced 50 kW. Using mathematical models and simulations, the researchers expect the plant to produce 320,000 kWh per year.

Though the concept seems simple, producing the technology to carry out the project was no small feat. The structures installed on board required advanced engineering to ensure none of the moving parts were in direct contact with salt water. Engineer and Project Manager Edgar Kvernevik explains that electricity is produced with the help of a fluctuating water column, a concept that has been demonstrated to be a trusted approach. The vessel can be left to swing its anchor in any part of the ocean with sufficient wave energy. A special anchoring system aligns the anchor so that the vessel always faces incoming waves to optimize efficiency. What is even better is that this can all be controlled remotely from onshore.

Currently tests are being conducting in an area of Norway where the high annual average wind speed provides a well-suited region for the exploitation of renewable energy from wind and waves. However development will not stop here. Once electricity generation is ensured, hydrogen production plants will be installed on the vessel so electricity can be stored in the form of hydrogen gas. Capacity can also be enlarged to 1 MW by installing modules on larger vessels. Finally to take advantage of both wind and wave energy, platforms have been designed to carry a 4 MW wave power plant with a 6 MW wind turbine installed on top.

SINTEF. "Fishing vessel transformed into a wave power plant." ScienceDaily. ScienceDaily, 9 February 2015. <www.sciencedaily.com/releases/2015/02/150209094940.htm>.

Portland Plans to Generate Electricity through Water Pipes
by Niti Nagar

Lucid Energy has created the LucidPipe Power System which harnesses the water flow in municipal pipelines to produce hydroelectric power. The LucidPipe is installed in a section of an existing gravity-fed conventional pipeline that is designated for transporting potable water. The water flows through four 42-inch turbines, each connected to a generator outside the pipe. In Portland, Oregon, the 200 kW system was privately funded by Harbourton Alternative Energy. Although the power system was installed in December, it is currently undergoing reliability and efficiency testing. So far, it has been reported that the presence of the turbines does not slow the water flow rate significantly, so there is no change on pipeline efficiency. The system is set to begin generating power at full capacity by March.

Once running, the system is expected to generate approximately 1,100 MWh of energy per year. This is equivalent to the amount of energy needed to power about 150 homes. It is projected that over the next 20 years, the system should generate about $2 million in energy sales to Portland General Electric. Harbourton Alternative Energy will get a share of these sales and it plans on sharing the money with the City of Portland and the Portland Water Bureau in order to offset operational costs. At the end of the 20-year period, the Portland Water Bureau will have the option to purchase the system, along with all the energy it produces.

Currently, this system is the only one of its kind in Portland. However if shown to be successful, more may follow. In Riverside, California a previously-installed energy system has been providing power since 2012. Since then many smaller, but similar, systems have become available, many of which can be installed within households. The Pluvia, for example, generates electricity from the flow of rainwater off of rooftops, while the H$_2$0 Power radio uses electricity generated by the flow of shower water.

Coxworth, Ben. "Portland to generate electricity within its own water pipes." GizMag, 17 February 2015. < http://www.gizmag.com/portland-lucidpipe-power-system/36130/>.

Difficulty of Harnessing the Ocean's Great Power

by Shannon O'Neill

There is great interest in harnessing the energy of the ocean's waves and tides, stemming from the ocean's ability to provide a constant energy source that does not depend on weather conditions and produces no hazardous waste— drawbacks that are often associated with other forms of renewable energy resources. The first large-scale tidal project, La Rance, opened in France over fifty years ago, and since, a similar project has been developed in Korea, and smaller projects developed in China, Canada, and Australia. Despite the great capacity to harness the energy of the ocean, the combined production of these projects produces only 0.5 GW of power versus wind energy's production of 400 GW.

Such tidal projects require areas with a great tidal range, meaning a large difference in the height of the water between low and high tides. A barrier is built that holds the water when the tide goes out. The process of releasing water back into the ocean drives turbines that harness this energy. These turbines are powered again when there is a high tide and water is let back in. Tidal projects have been greatly limited due to the fact that not many areas possess the large tidal range necessary to fuel the turbines. Additionally, the barriers are typically built across estuaries, causing great harm to these areas that contain high levels of biodiversity.

In 2008, the 1.2 MW Seageneration project was installed in Northern Ireland. This project harnessed the energy from currents rather than tidal range. It works as currents and the tide moving in and out drive a vertical axis turbine that is anchored to the seafloor. So far, extensive research has been conducted and it has been ensured that this project has provided no threats to the biodiversity of this area. Developing from this concept, the proposed MeyGen project set to take place in Scotland will use a traditional three-bladed, horizontal axis turbines that is proposed to produce 400 MW by the early 2020s.

Though such tidal projects and their new technology provide a good amount of energy, these projects rely heavily on areas where there are high ocean currents. Therefore, the ability to harness the energy from the ocean's waves provides an even greater potential energy source, as large waves occur in many more areas usually associated with high wind speed. Despite the 300kW Mutriku wave project in Spain, developers have had difficulty in creating a system that can withstand the harsh ocean conditions.

There has been great frustrating regarding the slow development of both tidal and wave power. However, such transformative technology will require time and money. Despite this slow progress, it is believed that tidal and wave power could provide the United Kingdom with 20% of its total energy needs.

BBC News: Riding the Waves: The challenge of harnessing ocean power
(http://www.bbc.com/news/business-31651019)

CorPower's Recent Breakthrough in Wave Energy Technology
by Melanie Paty

In a recent article posted on KTH, the Swedish Royal Institute of Technology, Peter Larsson reports on the wave energy technology breakthrough of CorPower Ocean. Wave energy technology has been limited by its cost, but CorPower Ocean, a Swedish company, has designed a gearbox system that generates four times more energy than competing systems at one third of the cost. The CorPower converter, developed in collaboration with KTH researchers, is a point absorber system, using a bobbing buoy to turn gears that drive the generator below the surface. The novelty of the recent development is its cascade gear model, a rack and pinion gear system in which the rack is moved up and down by the motion of the buoy, turning eight small gear wheels that evenly share the force, allowing the system to handle heavy loads and high velocities efficiently. This system easily converts rotational motion to linear motion and vice versa. Waves are unreliable given their variety of height and speed, so it has been difficult to develop a conversion system that works with all wave types; however, CorPower's CEO Patrik Möller claims that this converter is compatible with the entire spectrum of waves. The device won €100,000 at the MIT Building Global Innovators Demo Day.

Larsson, Peter. "Making waves with new gear technology. Researchers' Innovation reduces cost of producing wave energy." KTH. February 10 2015. (https://www.kth.se/en/aktuellt/nyheter/hjartat-inspirationskalla-for-nytt-kraftverk-1.529802)

Callahan, David. "What do gearboxes have to do with ocean energy? A lot." KTH. February 19 2015. (http://www.kth.se/blogs/stockholmtech/2015/02/what-do-gearboxes-have-to-do-with-ocean-energy-a-lot/)

Environmentalists Sue Governmental Agencies in an Effort to Help Pallid Sturgeon in Montana Rivers

by Trevor Smith

The Bozeman Daily Chronicle reports that two environmental activist groups have filed a lawsuit early this week against the US Army Corps of Engineers, the Fish and Wildlife Service, and the Bureau of Reclamation (Lundquist 2015). The Natural Resources Defense Council and the Defenders of Wildlife's suit claims that these agencies' operation of dams on the Montana and Yellowstone Rivers threatens the life of pallid sturgeon. The suit hopes both to stop the agencies' current actions, which it claims will be ineffective in helping the fish survive, and to force the agencies to create a new dam modification plan.

Pallid sturgeon have been listed as endangered since 1990, and although their population is estimated to have increased somewhat since then (Brown 2015), biologists assert that the upper Missouri River pallid sturgeon fish population rests at approximately 125 fish, almost all of whom are older—younger fish are not surviving (Lundquist 2015).

The problem comes from the way the two dams in question work. A study published by the American Fisheries Society in Fisheries last month makes the novel claim that one of the main reasons the dams threaten pallid sturgeon is not because of their difficulty passing through the dams, but because the dams slow the speed of the water, creating anoxic "dead zones" that lack enough oxygen for the fish to survive (Guy *et al.* 2015). The study is notable in that it focuses on the effects of dams on fish survival upriver of the dams, noting that dams make life more difficult for pallid sturgeon miles before they attempt to cross the dam.

The lawsuit cites this evidence to argue that the US Army Corps of Engineer's current plan to aid pallid sturgeon survival—increasing the width of side channels for fish to navigate through dams—is unlikely to be particularly effective at increasing the size of the sturgeon population (Brown 2015). The lawsuit seeks both to block this current plan and to require governmental agencies overseeing the dams to make different modifications to improve the health of the rivers for the pallid sturgeon.

Brown, Matthew. "Advocates: Dams Put Dinosaur-Like River Fish at Risk." ABC News. February 2, 2015. http://abcnews.go.com/Technology/wireStory/advocates-dams-put-dinosaur-river-fish-risk-28673490

Guy, Christopher S., Treanor, Hilary B., Kappenman, Kevin M., Scholl, Eric A., Ilgen, Jason E., Webb, Molly A. H. "Broadening the Regulated-River Management Paradigm: A Case Study of the Forgotten Dead Zone Hindering Pallid Sturgeon Recovery". Fisheries. http://news.fisheries.org/broadening-the-regulated-river-management-paradigm-a-case-study-of-the-forgotten-dead-zone-hindering-pallid-sturgeon-recovery/

Lundquist, Laura. "Groups sue to save endangered pallid sturgeon". The Bozeman Daily Chronicle. February 2, 2015. http://www.bozemandailychronicle.com/news/environment/groups-sue-to-save-endangered-pallid-sturgeon/article_38667c30-9954-5d29-83c2-18a6bff32634.html

Proposed Tidal Lagoon Power Plant Could Provide Enough Energy to Power all of Wales

by Trevor Smith

Tidal Lagoon Power has unveiled its proposal for what would be the world's first tidal lagoon power plants. A single plant operating off the coast of Cardiff could supply enough energy to power all of Wales; the company ultimately hopes to build six plants along the coast of the United Kingdom (Gosden 2015). The plants, which depend on predictable tide movements, are being hailed as much more reliable than other clean energy sources like sun or wind power; however, environmentalists have voiced concerns about their impact on sea wildlife.

The power plants rely on building extensive sea walls around tidal lagoons. Gates in the walls would allow free movement of water except when the tide is rising or retreating. Before the tide begins to rise, the gates are shut and water builds up and pushes against the outer side of the sea wall. When high tide is reached, the gates are opened; the movement of water into the lagoon turns turbines that produce electricity. A similar process occurs as the tide falls. The gates are shut before the tide retreats, then are opened when low tide occurs, again causing a rush of water that turns turbines (Harrabin 2015). Since tides rise and fall at predictable daily patterns, the power produced from the turbines is consistent and reliable, unlike generators that rely on sunny or windy weather to produce energy.

The plants have been met with some opposition, however. Environmentalists have questioned the plants' impact on fish migration between nearby rivers and the lagoons, and are concerned about the disruption on estuaries and natural tidal movements the plants will cause (Harrabin 2015). Tidal Lagoon Power largely dismisses these claims, arguing that the sea walls will function as human-made reefs.

If approved, tidal lagoon power could be supplying a significant percentage of UK power by 2022.

Gosden, Emily. "Cardiff tidal energy lagoon 'could power every home in Wales'". The Telegraph, March 2, 2015.
http://www.telegraph.co.uk/news/earth/energy/11445579/Cardiff-tidal-energy-lagoon-could-power-every-home-in-Wales.html
Harrabin, Roger. "World's first lagoon power plants unveiled in UK". BBC News, March 2, 2015. http://www.bbc.com/news/science-environment-31682529

Making Big Waves in Alternative Energies Energy
by Abigail Wang

For over 40 years countries in Europe and the United States have invested millions to harness wave and tidal power. Recent efforts to reduce fossil fuels have spurred efforts to look closer at technological advancements for tidal energy. So even though the sustainability of wave and tidal energy is still questionable, it has a promising future.

The US Department of Energy's Water Power Program has recently focused more on new technologies that use marine resources while achieving economic viability. The growth in this sector in 2015 will likely include technological advancements for prototypes and early production models; commercial deployments will only happen once the technological costs become more competitive with local costs of energy. The cost for tidal wave technologies is still extremely high, but as the technology continues to develop it's likely that these expenses will decrease.

The US Department of Energy has already completed three of its six 2014–2015 wave and tidal energy projects. Project M3 Wave determined the commercial viability of the Delos-Reyes Morrow Pressure Device, which harnesses the oscillating motions of waves by inflating and deflating bags to convert wave energy into electricity. Another project expected to deploy by the end of 2015 is the Roosevelt Island Tidal Energy Project (RITE) where 30 turbines will be installed into New York City's East River to provide 1,050 kW of energy to 9,500 residents.

According to Georgia Tech Research Corp, as of 2011 the maximum power potential for US tidal energy is 50,783 megawatts. The International Renewable Energy Agency (IRENA) estimates that the potential for tidal energy worldwide is 1 terawatt.

There are currently two methods of tidal wave energy conversion: tidal range and tidal current technologies, both of which use turbine-spinning generators. The former uses dam-like structures and height differences between low and high tides to create streams of high-velocity water flow. The latter places turbines under water in the path of strong tidal currents and converts the kinetic energy of these currents into electricity.

The oldest operating tidal range dam, La Rance Power Plant, is located in the Rance River in Brittany, France. The site has 24 turbines that produces 6,000 GWh per year and provides 90% of Brittany's electrical needs. A similar project dubbed POWER (Partners Offering a Water Energy Revolution) program is being developed off the east coast of China. It will use dynamic tidal power by constructing a dam running perpendicular to the shore, interrupting tide current and capturing the water flow from both sides of the dam. This program is planned to generate 15 GW of electrical energy and could generate up to 23 billion kWh per year.

The Tidal Energy Ltd. (TEL) in Wales has planned two projects that will use DeltaStream technology. This technology involves connecting three 50-ft diameter turbines using a 400-ton triangular frame. As water velocity increases, the turbines will operate at the maximum power coefficient, which is the ratio of power extracted to the total power available in water currents.

What happens to the surrounding environments where the turbines are installed? La Rance Power Plant saw a decrease in animal life during its initial implementation, but the populations of cuttlefish and sea bass have slowly returned since. Newer turbines are reportedly 'fish-friendly,' do not kill or disturb aquatic life, and have less impact on the environment. Additionally, sites are now required to pass an Environmental Impact Assessment and are closely monitored to deal with any unpredictable problems.

As countries continue to look for energy alternatives to lower their fossil fuel consumption, more money will be invested into technologies like tidal-wave energy.

Williams, Andrew. "Wave and Tidal Energy in 2015: Finally Emerging from the Labs." Renewable Energy World: 25 February 2015.
http://www.renewableenergyworld.com/rea/news/article/2015/02/wave-and-tidal-energy-in-2015-finally-emerging-from-the-labs

Gonzalez, Carlos. "For an Energy Source, Look to the Sea." Machine Design: 26 February 2015. http://machinedesign.com/energy/energy-source-look-sea

WIND POWER

Conventional wind turbine technology has gradually standardized to blade sizes that can be moved by big rigs on highways, resulting in vast turbine arrays spread out across the landscape in many developed countries. Where there is enough wind over land or just offshore, and there is public acceptance of the appearance of these arrays, wind is a source of power rapidly becoming price-competitive with fossil fuel generation. But there remain substantial untapped resources in developing countries, and farther offshore, and higher in the sky, where the winds are more reliable and stronger.

Kenyan Government Solves Energy Issues with Largest Wind Power Project in Africa
by Jessie Capper

Unreliable, inefficient, and expensive electricity is a continuous issue in third-world countries, especially Kenya. Kenya's energy consumption increased by 9% between 2010 and 2011, and demand is expected to grow a further 12% by 2030 (Court Jan 29, 2015). This expanding demand for energy has presented numerous obstacles for the Kenyan government and Kenyan power companies. Many Kenyan communities face costly energy bills and recurring interruptions in power supplies; these widespread interruptions affected 75% of the country in 2014. In January 2015 alone, the Kenyan Power Company—the country's main electricity transmission company—recorded roughly 9 energy interruptions for every 1,000 customer at the household level. As a result, the Kenya government is trying to reduce its dependence on hydropower—which provides 65% of the country's electricity—due to Kenya's unreliable rainfall patterns. As these trends persist, electricity and power companies, along with Kenyan government officials, are developing a reliable, cost-effective, and renewable energy source for Kenyan communities (Court Jan 29, 2015).

The country has commenced one of its most ambitious wind energy projects, predicted to add 5,000 MW of power to the Kenyan energy grid. The 300 MW Lake Turkana Wind Power Project is set to produce 20% of the country's current electricity generating capacity once completed in 2016. The main goals of the Wind Power Project are to provide reliable, low-cost wind power to the national grid. It includes 365 wind turbines, each with a 52-meter blade span, and a capacity of 850 kW (Lake Turkana Wind Power).

Unfortunately, plans for past power plants to improve Kenya's energy supply have failed due to a lack of bids from construction companies. The development of the Dongo Kundu and Liquid Natural Gas facility—which was set to generate 5,000 MW of electricity—was halted due to its expensive construction cost of $1.44 billion (IPP Journal Sept 8, 2014). Although the Lake Turkana Wind Power Project is promising for the country's energy supply, it is also a costly project amounting to $694 million—establishing the largest private investment in Kenyan history. However, the procurement process for the Lake Turkana Wind Project has already proven to be much more successful than the Dongo Kundu and Liquid Natural Gas project. An international collaboration among lenders and producers—including the African Development Bank and the British company Aldwych International and Standard Bank—have worked together to pay for and install the 365 wind turbines. Once developed and in operation, the ambitious Lake Turkana Wind Power Project will be the largest wind farm on the African continent and a hopefully replicable solution for other countries to follow.

Court, Alex. "Will Africa's biggest wind power project transform Kenya's growth?" CNN. January 29, 2015. Accessed February 18, 2015. http://www.cnn.com/2015/01/29/business/ltwp-kenya-windpower/
Lake Turkana Wind Power (http://www.ltwp.co.ke/the-project/overview)
"Kenya to re-tender 700 MW LNG facility." IPPJournal. September 8, 2014. Accessed February 18, 2015. http://ippjournal.com/2014/09/kenya-to-re-tender-700-mw-lng-facility/

The Benefits of Offshore Wind Energy
by Alex Elder

In the past decade, the use of wind energy has increased dramatically. Wind farms that provide energy for millions of homes have popped up all over the country. However, the current energy market still relies heavily on the production of oil and gas resources as a main source of energy. The two biggest downsides of fossil fuel dependency are the lack of renewability and the environmental and economic consequences. For these reasons, renewable energy sources like wind power have become more appealing in recent years as the negative impacts of using fossil fuels become more salient and serious. In particular, offshore oil rigs are one of the most controversial sources of fossil fuel because they are notoriously dangerous for those working on them and result in frequent oil spills and fires, which negatively impact the ocean through pollution.

Rather than relying on offshore oil rigs, a different offshore alternative is the installation of wind turbines off the coast. Although Europe has already established many successful offshore wind farms, the idea has yet to catch on in other parts of the world like the United States. Offshore wind farms are more beneficial than oilrigs in many ways. First of all, they are much safer to work on than offshore oil rigs because there is little risk of fires or the use of dangerous equipment. Because they are also offshore, transfer of jobs from oil rigs to offshore wind farms. Secondly, because these farms would be harnessing wind power rather than drilling for oil, there is no environmental risk like a potential oil spill or fire. Furthermore, because wind is a renewable energy source, offshore wind farms could continue to provide safe and stable jobs and stimulate the economy in a more reliable manner.

Not only are offshore wind farms a better alternative than offshore oil rigs in terms of risk, cost, and renewability, they also have significant benefits over their onshore counterparts. Unlike onshore wind turbines, offshore installations are less obtrusive both in terms of noise and aesthetic disruption of the landscape. Moreover, wind speeds over water tend to be much higher than on land. Thus, offshore wind turbines would yield more power overall than onshore turbines.

Overall, offshore wind farms could be a great alternative energy source to both offshore oil rigs as well as onshore wind farms. Although their use is popular in some parts of the world, they have yet to be used in the United States. Future use of offshore wind farms could reduce negative environmental impacts and stimulate the economy all while providing renewable, clean energy.

Snyder, B., and Kaiser, M. "Ecological and economic cost-benefit analysis of offshore wind energy." Renewable Energy 34.6 (2009): 1567-1578.

Menaquale, Andrew. Oceana. "Offshore Energy by the Numbers" (2015) http://usa.oceana.org/sites/default/files/offshore_energy_by_the_numbers_report_final.pdf

Energy Kites: Airborne Wind Turbines
by Alex Elder

Makani Power, acquired by Google X in 2013, is seeking to improve the modern wind turbine design in order to make wind power more efficient, cheaper, and less intrusive. Makani has developed an innovative new system of utilizing wind as a source of energy by using an "energy kite." These kites are actually carbon fiber gliders which fly in circles while remaining tethered to the ground. This design allows the energy kites to reach higher altitudes than traditional wind turbines while significantly reducing the cost of materials for construction. Due to their lightweight design, the kites are more aerodynamic, and thus more energy efficient, than ground-based wind turbines. Because higher altitudes allow access to stronger and more consistent wind speeds, the kites can generate about 50% more energy than traditional wind energy technology.

The kites are first launched from the ground station by rotors which act like propellers on a helicopter. Once in the air, the kite generates power by flying in large circles where the wind is strongest. Air moving across the rotors mounted on the kite forces them to rotate, driving a generator to produce electricity. This power then travels down the kite's tether to the grid below. The tether itself is made of conductive wires and connects the kite to a ground station. In 2013, Makani Power demonstrated that these energy kites can even complete the launching, circling, and landing process entirely autonomously, meaning that the system is generally low-maintenance as well.

These energy kites can even operate over the ocean, where winds are stronger than on land. Having an energy kite tethered to a floating object like a buoy or a boat would solve many of the problems associated with off-shore wind farms which must be mounted on the sea floor. Overall, this innovative idea opens the door to a more energy-efficient and cost-effective system of harvesting wind energy.

Anderson, R. (2015). Wind turbines take to the skies to seek out more power. BBC News. http://www.bbc.com/news/business-31300982
Makani Power. http://www.google.com/makani/solution/

Smart Source with Skystream
by Alexander Flores

Xzeres, a small corporation based in Wilsonville, Oregon has developed a cost-efficient, compact, and all-inclusive personal wind generator designed to function in very low winds

(10–12+ mph). This personalized wind turbine known as the Skystream has a height of 52 feet, a weight of 170 pounds, a rotor diameter of twelve feet, a swept area of 116 feet, three reinforced composite fiberglass blades, a slot-less, brushless permanent magnet alternator, a battery charge controller kit, a braking system with electric stall regulation, and comes with Skyview monitoring software. This particular software allows one to simply track the Skystream's performance directly from a personal computer. Installation takes approximately two days including installing the foundation, erecting the turbine, and connecting it to the meter. Essentially, when the wind blows, the turbine generates clean, affordable energy (400 kilowatt hours per month) to power a home or business. If more energy were to be used than the turbine is producing, utility power takes over. If less energy were used, the excess electricity could be sold back to the electric company. Skystream turbines can also be installed to facilitate a variety of grids, plants, gas stations, schools, and can even be placed on light posts in public spaces. One could even supplement energy production with the utilization of the Skystream in conjunction with solar panels or other electric generation technologies. It shouldn't be long before we begin to see more products like the Skystream in public spaces, making energy-savings a breeze.

Xzeres Wind 1 (http://www.windenergy.com/products/skystream)
Xzeres Wind 2 (http://www.windenergy.com/products/skystream/skystream-3.7)
Xzeres Wind 3 (http://www.windenergy.com/markets/utility-connected-homes)
Xzeres Wind 4 (http://www.windenergy.com/markets/business-retail)

Don't Like Turbines? Try Flying a Kite
by Alison Kibe

Recognizing that as much as 90% of a wind turbine's power comes from force exerted on the tips of turbine blades, David North, an engineer at NASA's Langley Research center, asked, "What if I had a machine that was just the tip of the blade?" (Silberg, 2012). North began developing a kite prototype to replace the 400-ton turbine tower to better understand the aerodynamics that would be involved (NASA, 2012).

It is the simplicity of an energy-generating kite was appealing to David Shaefer who is now the CEO and founder of an Oregon based startup eWind Solutions that is working to build a commercially useful Airborne Wind Energy (AWE) system (Owen, 2015). Unlike a traditional wind turbine, the AWE system, or energy-generating kite, could reach higher altitudes where wind is faster and steadier (eWind Solutions).

The AWE will also use less material and as a result cost less and may have smaller environmental impacts (eWind Solutions).

According to eWind, the AWE Schaefer and his team are building works by using a specially designed kite that the wind blows in a continuous figure-eight motion. This motion moves the kite string that applies force to a drum or wheel attached to a generator. eWind kites will also use sensors to determine optimal flight paths and indicate when the kite string should be extended or reeled in based on wind conditions. Ideally, the kite should be able to fly under wind speeds as low as four miles per hour and generate four times more electricity than a traditional turbine (Owen, 2015).

The idea has struck a chord with others, including the US Department of Agriculture, which has awarded eWind Solutions a $100,000 grant and potentially $1 million more (Owen, 2015). The eWind Solutions team now hopes to create a prototype that will be ready for commercial use sometime next summer (Owen, 2015).

eWind Solutions. Technology. 2014. http://www.ewindsolutions.com/technology/

Owen, Wendy. Wind energy: Beaverton startup builds kites to replace turbines on farms. The Oregonian. April 2015.
http://www.oregonlive.com/beaverton/index.ssf/2015/04/beaverton_wind_energy_startup.html

Silberg, Bob. Energy Innovations: Electricity in the Air. NASA: Global Climate Change. May 2012. http://climate.nasa.gov/news/727/

Beefing up a Wind Turbine with Compressed Air

by Emil Morhardt

If your wind turbine isn't going fast enough to meet the demands of the grid, blow on it a little harder: that's the general idea suggested by Sun *et al.* (2014). The concept is a little like a hybrid electric vehicle; if the internal combustion engine isn't going fast enough, give it a little boost from the electric motor connected to it. Except in this case, it's that if the wind turbine isn't going fast enough, goose it with a little compressed air. You might be envisioning a compressed air nozzle pointed at the turbine blades, but there's a better way: use a motor driven by compressed air to speed up the turbine. One novel aspect to this study is that the device envisioned as a compressed-air motor is something called a scroll expander, or scroll-type air motor, a new type of pneumatic drive, but that doesn't seem to be central to the idea—any suitable air-driven motor should work. The main point is to have it integrated with the wind turbine so that when needed, it can help out in the short term.

To function, it needs a source of compressed air, but where that might come from isn't dealt with in the paper. It could be generated by the same wind turbine using excess electricity to run an air compressor, but maybe there are other sources as well, and I think the authors envision a centralized source of it— rather than one compressor per wind turbine, one compressor per field of turbines, all connected with air hoses. The authors created a test rig in the lab that simulated all aspect of the system, and conclude that 55% of the energy used to compress the air could be expected to be returned from the boosted wind turbine, which is fairly good for pneumatic motors, and the wind turbine itself was much better at providing steady reliable power with the compressed air device than without.

Sun, H., Luo, X., Wang, J., 2014. Feasibility study of a hybrid wind turbine system–Integration with compressed air energy storage. Applied Energy DOI: 10.1016/j.apenergy.2014.06.083.

Mini Piezoelectric Wind Turbine
by Emil Morhardt

In the Internet-of-Things, distributed sensors that collect data and transmit it to each other (and ultimately to a computer somewhere) wirelessly will become commonplace. They need a power source, but don't want wires, or batteries that can wear out. Small photovoltaic panels work in some instances, but not always. Hence wouldn't it be nice if a very small wind turbine came along. Yang *et al.* (2014), writing in Applied Physics Letters, invented an interesting one. It consists of a wind powered rotating drum, 3 cm in diameter, with several elastic balls inside that get dumped onto piezoelectric cantilevers as the drum rotates: the energy of the wind gets transferred to potential energy in the form of lifted balls, which then transfer their energy to the piezoelectrics when they fall on them. The piezoelectrics convert the kinetic energy of the balls into electricity, which can then be stored in a small supercapacitor for use as needed. No wires, no batteries, and not much mechanical that can wear out. Plus, the parts are cheap and readily available. And cleverly, the ends of the piezoelectric levers sticking out around the edge of the drum make perfect wind catchers.

Yang, Y., Shen, Q., Jin, J., Wang, Y., Qian, W., Yuan, D., 2014. Rotational piezoelectric wind energy harvesting using impact-induced resonance. Applied Physics Letters 105, 053901. Photo from the paper. Link to the abstract: http://bit.ly/V9aYEC

Solar and Wind Producers in Japan, Portugal, Design Floating Power Plants

by Trevor Smith

As swathes of land large enough to house increasingly sizable wind and solar energy plants become harder to find in densely populated countries, floating power plants have emerged as a latest trend in renewable energy production. Two countries in particular have taken an interest in utilizing floating technology: Japanese company Kyocera TCL Solar recently finished a large floating solar plant with plans in place to construct an even larger one, and a Portuguese scheme to build a nexus of floating wind turbines has just been approved by the European Commission on energy.

Japan's high population density creates issues for solar energy production, which has traditionally required large open areas of land for solar panels to be laid down. Kyocera TCL Solar looked instead to Japan's many lakes, which provide an excellent site for solar power in part because of their proximity to Japan's cities (Upadhaya (1) 2015). But the benefits for aquatic solar panels expand beyond site availability. Traditionally, solar panels have been placed in hot areas where they heat up dust can partially cover solar panels, reducing their efficiency (Upadhaya (2) 2015). The water the panels float on provides a natural cooling process for the panels, boosting efficiency since photovoltaic solar panels are most efficient when they are cool, and energy would otherwise have to be used to artificially cool the panels. The moist environment of the lakes also sidesteps a cleaning issue through their position in a relatively dust-free environment.

Meanwhile, Portugal seeks to install floating wind turbines not in lakes within its borders, but in the ocean just off its coast. Although wind turbines planted in water are nothing particularly new, the technology Portugal hopes to rely upon will allow the placement of turbines in much deeper water than previously possible. Turbines placed in bodies of water have typically been attached to either the lake or ocean floor. The WindFloat technology Portugal is using for this project allows the turbines to float in place offshore, opening considerable amounts of new water for the possibility of wind energy production.

Hill, Joshua S. "EU Commission Approves Portuguese Floating Wind Farm". Clean Technica. April 27, 2015. http://cleantechnica.com/2015/04/27/eu-commission-approves-portuguese-floating-wind-farm/

Upadhyay, Anand (1). "Kyocera TCL Solar Completes Two Floating Solar Power Plants In Japan". Clean Technica, April 27, 2015.

http://cleantechnica.com/2015/04/27/kyocera-tcl-solar-completes-two-floating-solar-power-plants-japan/

Upadhyay, Anand (2). "Why the Buzz Around Floating Solar PV?" Clean Technica, April 13, 2015. http://cleantechnica.com/2015/04/13/buzz-around-floating-solar-pv/

GEOTHERMAL ENERGY

Hot springs and geysers in tectonically active areas of the globe are an obvious source of energy, and many such sites have been fitted with steam collection facilities to drive conventional steam turbines. Others, with lower level heat bring hot water to the surface to vaporize organic working fluids with lower boiling points, which also drive turbines. In developments reminiscent of the hydraulic fracturing for oil and gas, there are new pushes toward less obvious sources of geothermal energy, deeper wells, some expanded using processes analogous to fracking, and some using non-conventional fluids. If these work out, there may be much more geothermal energy available than was recently thought.

Iceland's Deep Drilling Geothermal Energy Project

by Shannon O'Neill

Iceland's unique geology has made it a prime region for the development of geothermal energy. Specifically, the Reykjane Peninsula, located on the Mid-Atlantic Ridge on the southwestern coast and the home of four volcanoes, is a prime region for such development. Its volcanic geology provides geothermal pools that are heated by the steam and magma deep below the service. Geothermal wells harvest the heat from the pools to power turbines, providing one hundred megawatts of power, enough to power thousands of homes in the region. Iceland is powered almost solely on renewable energy resources, with geothermal energy contributing to a fourth of such resources.

Scientists are hoping to take geothermal energy to the next step with the Iceland Deep Drilling Project (IDDP). This will project will involve drilling a geothermal well into Reykjane's volcanic field, thousands of feet below the surface, in order to harvest energy from both the superheated steam and the molten rock. This will provide an unconventional resource to the geothermal power generators, as the wells would tap into "supercritical" water deposits, where the fluids are under such immense pressure and heat that they are in state that is neither liquid nor gas.

Harnessing these supercritical deposits has the potential to make Iceland the world's leading geothermal energy exporter. However, the issue lies with dealing with this extremely hot supercritical material. The supercritical water deposit was first discovered five years ago by accident. A special steel casing was placed in the well, keeping it intact for two years. Until a valve broke, the superheated steam from the well provided more than half of a power plant's 60-megawatt output.

Ever since, scientists have been trying to further develop supercritical wells to harness the massive energy potential in these deposits. This would not only increase the output in Iceland, but also in other areas that have high geothermal activity, providing more renewable energy resources. There has been some concern that such wells can induce earthquakes and possibly even volcanic eruptions, however geologists have not supported such claims. The technology is there, rather it is a matter of wells becoming more price-competitive to nonrenewable resources such as oil.

CNBC News: Are volcanoes the energy source of the future?
(http://www.cnbc.com/id/102261363)

Renewable Energy Resource Revolution in Nicaragua
by Shannon O'Neill

Nicaragua's intense winds, sunny weather, and volcanoes, provide multiple renewable energy resources and the capacity to generate 5,800 MW in electricity annually. However, until recently, these resources have gone virtually untapped. In 2005, the government began implementing technology to harness these valuable resources. Currently, renewable energy is already contributing to half of the country's electricity, and government officials believe that this will rise to 80% in the next few years. This is staggering, as the US only receives 13% of its electricity from renewable resources.

Such advancements are particularly important for Nicaragua, as the country has no oil of its own, forcing it to rely heavily on imported gas oil for its energy in the past. This expense along with their lack of thermal plants necessary for turning the oil into electricity left the country with rolling twelve-hour blackouts that have had drastic negative impacts on the economy.

Policies, such as tax breaks, were enacted along with the welcoming of private investments in order to promote the development of renewable energy projects. Currently, Nicaragua is the home to a massive wind farm, in an area that is claimed to be the best place for wind energy due to constant wind flow. This farm contributes more than forty megawatts of energy to the national network. Additionally, there has been great interest in the available geothermal energy. A Nevada-based firm has contributed more than $400 million dollars to construct a geothermal plant next to a volcano. Water that is heated by the molten rock is brought to the surface and then used to fuel a turbine and create electricity.

Though the developments in Nicaragua have already contributed to half of the country's electricity, only 5% of the renewable energy potential has been developed. In the upcoming years, there will be more development of hydroelectric dams, and more geothermal plants and wind farms to fully take advantage of the available resources.

"Nicaragua's Renewable Energy Revolution Picks Up Steam"
(http://www.npr.org/blogs/parallels/2015/03/11/392111931/nicaraguas-renewable-energy-revolution-picks-up-steam)
"Nicaragua: a renewable energy paradise in Central America"
(http://www.worldbank.org/en/news/feature/2013/10/25/energias-renovables-nicaragua

AltaRock's Enhanced Geothermal Systems
by Ali Siddiqui

In Bloomberg magazine, an exciting article by Adam Aston, Pete Engardio, and Joel Makower (Executive Editor, Greenbiz.com) provides a list of 25 companies to watch in the energy technology sector. Their list is interesting because its selection criteria made sure to include companies that were actually selling their newly innovative products in the market, were not publicly listed, often unheard of, and had large sums of financial backing from high profile venture capitalists or companies. Within this list was one particularly fascinating company named AltaRock Energy, Inc. founded and headquartered in Seattle, Washington, whose new approach to extracting geothermal energy may change the landscape of the energy sector.

AltaRock's goals are to not only promote the increasing use of geothermal energy by building geothermal energy plants, but also to use it's team, technology, and approach to turn under performing assets into high profitable energy projects. They are able to accomplish these goals with an interesting new technology known as Enhanced Geothermal Systems (EGS). Currently natural extraction of geothermal power requires the use of high temperature permeable rock with naturally occurring water circulation at depth, which is not only finite and rare, but also carries a heavy cost and high risk for failure when drilling begins. AltaRock's EGS create artificial geothermal reservoirs by using hydraulic pressure to create a small network of fractures in the rock below that allow the rock to act as a radiator increasing permeability and circulation. This will make geothermal extraction less costly, more successful, and more available. While this sounds similar to fracking, the difference stems from the pressure used to fracture the rock. EGS maintains a lower pressure by utilizing a process known as hydroshearing. They are able to seal fractured zones from one another in order to maintain the network, as well as maximize circulation and heat, providing them a superior system for extracting energy.

This company is currently managing over 40,000 acres of land located on the western bank of the Newberry Volcano in Central Oregon administered by the US Bureau of Land Management in order to carry out studies for how successful their system will be. Smart energy investors should pay attention.

Aston, Adam, Engardio, Peter, Makower, Joel. 25 Companies to watch in Energy Tech. Bloomberg. January 26th 2017.
http://www.bloomberg.com/ss/09/07/0714_sustainable_planet/1.htm
Altarock.com (http://altarockenergy.com/technology/enhanced-geothermal-systems/).

The United Forest Service Hopes to Lease Washington National Forest for Geothermal Development

by Trevor Smith

Over 80,000 acres of public forest land in the Pacific Northwest of the United States could soon be open to bids from geothermal energy companies. Washington is home to a considerable amount of geothermal activity, as indicated in the many volcanoes dotting its mountain ranges. Despite recognition that the Northwest is likely only a moderately fertile site for geothermal power production and potential environmental concerns, the decision to open the land for lease to private contractors is expected to go through by the end of the month (IJPR 2015).

The land comes entirely from Washington's Mt. Baker-Snoqualmie National Forest. The US Forest Service began to explore these plans after local utility companies expressed interest in the possibility of using parts of the land to produce clean, geothermal energy. The Snohomish Public Utility District, for example, hopes to build a geothermal plant on the land that could power nearly 20,000 homes (Ahearn 2015), a small but significant fraction of the 332,000 customers they currently serve (Snohomish County PUD 2015).

The Forest Service promises to weigh the possible energy and monetary benefits for the area against their potential environmental impacts. The plan requires any geothermal proposals to undergo a strict environmental review (IJPR 2015).

Washington currently lags behind other western states in geothermal energy. California is the worldwide leader for environmental energy, for example, and Oregon, Nevada, and Utah have all established some amount of geothermal energy. Washington actually has slightly lower geothermal potential than these states overall, and what land could serve as a good source of geothermal energy is protected. In other words, if Washington hopes to gain access to any sort of geothermal energy, it will probably have to come from making some public lands available for private geothermal production.

Ahearn, Ashley. "In Northwest, A Push To Protect Forest As Geothermal Projects Near" NPR. April 20, 2015. http://www.npr.org/2015/04/20/401039759/in-northwest-a-push-to-protect-forest-as-geothermal-projects-near

IJPR. "NW Forestland Could Be Leased For Geothermal Development." Jefferson Public Radio. http://www.npr.org/2015/04/20/401039759/in-northwest-a-push-to-protect-forest-as-geothermal-projects-near

Snohomish PUD. "About Us." Snohomish Public Utility District. 2015. https://www.snopud.com/AboutUs.ashx?p=1106

New Fluid Promises to Increase Efficacy of Enhanced Geothermal Systems

by Trevor Smith

A fluid out of the Department of Energy's Pacific Northwest National Laboratory promises to increase the efficiency of enhanced geothermal systems. PNNL scientists claim that less of their liquid, which is a mixture of water and polyallylamine, would be required to extract geothermal heat than the current mixture of water mixed with a few toxic chemicals. The new fluid has the added benefit of non-toxicity. PNNL notes that while the liquid was developed to increase the efficiency of geothermal energy production, it could also make oil and gas fracking more environmentally and human-health friendly than it currently is.

Enhanced geothermal energy relies on the same basic principles as traditional geothermal energy—heat in the ground is released and funneled in order to power turbines that create electricity. Where it works differently is how that heat is released. Traditional geothermal energy typically relies on existing heat sources that are either near the surface of the earth or that can at least be drilled down to easily. Enhanced geothermal energy seeks to access even heat that is trapped beneath rock layers that can't be easily drilled through. Instead, massive amounts of fluids are pumped down to create tiny cracks in rock layers that allow heat to escape. The new fluid out of PNNL dramatically decreases the amount of fluid required. After pumping their fluid down the rock layer, pressurized carbon dioxide is injected that bonds with the polyallylamine in the fluid, expanding the fluid to two and a half times its original volume and creating fractures in the rocks (EIN News 2015).

The fluid also holds promise for oil and natural gas fracking. Currently, fracking causes potential environmental damage because of the toxic liquids used to fracture the rock, in some cases, making the entire habitat around fracking sites nearly uninhabitable, with poor effects on the health of nearby residents (RT 2014). PNNL stresses that their polyallylamine mixture has not been explicitly tested for oil and gas extraction yet, but notes that in theory it ought to accomplish the task of current fracking methods without the health risks (EIN News 2015).

EIN News. "PACKING HEAT: NEW FLUID MAKES UNTAPPED GEOTHERMAL ENERGY CLEANER." Accessed 4/28/2015.
    https://www.einnews.com/pr_news/260256876/packing-heat-new-fluid-makes-untapped-geothermal-energy-cleaner
RT. "Living near fracking sites deteriorates health – study." 9/12/2014.
    http://rt.com/usa/187420-fracking-health-effects-pennsylvania/

NUCLEAR POWER

Is nuclear energy dead? Did the disaster at Fukushima-Daiichi kill it off? Not so fast. Even if fission of uranium in conventional power plants is more frightening than before, there are other versions of nuclear power in the works. Here are a few of them.

Laser Power Systems
by Dylan Goodman

Laser Power Systems (LPS) is a Massachusetts-based energy company founded in 2007 dedicated to creating a sustainable energy future through the use of new energy technologies based on thorium. Their goal is to create an alternative energy source to conventional nuclear power production. Thorium is a radioactive actinide metal that is abundantly available, but is currently fairly worthless. The federal government buried 3,200 tons of thorium in the Nevada desert simply because there was a lack of demand. (LPS) The problem with thorium is not a lack of abundance, rather the technology isn't available to use it to create power.

Thorium has not been largely considered as a useful fuel for future energy production. However, the value and use of thorium is not greatly understood. Laser Power Systems, however, has spent over 20 years researching and developing thorium-powered lasers and is finally planning on making their research public. Over the past decade, computer programs have allowed for a deeper analysis regarding differences in thorium and uranium 235, the type of uranium used in currently existing nuclear reactors. Conventional nuclear reactors, like many power plants, use heat to produce steam to turn a turbine. Laser Power Systems has created high-energy lasers, which are powered by thorium and used to heat water, produce steam, and turn a turbine. This process is extremely efficient; thorium contains 20 million times as much energy as a similar sized amount of coal. This allows thorium to be used as a long-term energy solution.

While a large-scale application of thorium power does not yet exist, Laser Power Systems has created a concept car that runs off energy produced by thorium. Clearly, the auto industry is not ready for thorium-powered engines, and that is not ultimately LPS's end goal. Ideally, this technology will be developed to be able to power the world. However, like many nuclear-related programs, the issue of radioactivity raises some concern.

Daileda, Colin. "One Day, Thorium Could Power Everything." Mashable. N.p., 07 Nov. 2013. Web.

Generating Power from Nuclear Wastes
by Alison Kibe

Nuclear waste from energy production hangs around for thousands of years and has driven many arguments against using nuclear power. However, others see nuclear waste differently, including Transatomic Power and Atlas Energy Systems who want to find a use for the million barrels of waste in the world (Henry, 2014). Transatomic Power's co-founder Leslie Dewan, sees tremendous potential for electricity production (TransatomicPower.com). "What we call nuclear waste isn't actually waste at all, because it has a tremendous amount of energy left in it," Dewan said in a Solve for X presentation in 2014, "That's why it's dangerous."

To use leftover energy, Transatomic Power is bringing molten salt reactors into the 21st century. In light water reactors used in today's nuclear power plants, fission products become trapped in the solid metal framework that contains the fuel. This buildup, in addition to radiation damages incurred by the reactor, limits how long the fuel lasts in the reactor and ceases the nuclear reaction Invented in the mid twentieth century, salt reactors use fuel dissolved in liquid salt that is less prone to damage and allows easy removal of fission products This means that fuels, new and spent, produce more fission reactions and can be used for much longer. The reason they have not been used in the past is because old molten salt reactors required near weapons grade enriched fuel to be utilized, were large, and didn't put out that much power per volume of salt solution. These new reactors use a liquid uranium-fluoride salt that allows the use of low-enriched uranium or spent nuclear waste (Transatomic Power, 2014).

Graphite moderators, which increase the likelihood of fission, were switched out for zirconium hydride that allows the reactor to be cheaper and more efficient by creating a more compact and power-dense system.

What is also impressive about a liquid reactor is that it does not require electrically sourced cooling mechanisms, the failure of which has led to the nuclear disasters in the past century. In the case of a failure, an electrically cooled salt plug melts and causes the salt solution to fill an auxiliary tank away from the moderators. A few hours later, the liquid salt freezes by itself, making it very safe in the case of an electrical failure.

Atlas Energy Systems wants to use nuclear wastes to ionize gases in order to create electricity (Henry, 2014). This system uses the reaction's kinetic energy, rather than the heat production associated with nuclear fission, to produce electricity. And, this reaction takes place in a power cell, so the nuclear waste could be portable or used off of the electrical grid (istart.com). The cell could last as long as the spent fuel remains radioactive, as this is what drives the electricity production (Henry, 2014).

Overall, the use of using nuclear wastes is promising and could be impactful. Transatomic power claims that if the 96% of remaining energy that exists in nuclear waste could be used, we could meet the world's energy needs for the next 72 years (TransatomicPower.com).

Henry, Hillary. Purdue students commercialize device that could convert nuclear waste into electricity. Purdue Research Foundation. December 3 2014.
http://prf.org/news/purdue-students-commercialize-device-could-convert-nuclear-waste-electricity
istart.org
http://istart.org/startup-idea/green-materials-energy/atlas-energy-systems/20300
TransatomicPower.com
http://transatomicpower.com/company.php
Transatomic Power. White Paper. March 2014.
http://transatomicpower.com/white_papers/TAP_White_Paper.pdf

Scientists at the National Ignition Facility Bring Us One Step Closer to Fusion Power
by Niti Nagar

Nuclear fusion seems to answer many concerns that we face with finding new sources of energy. Energy from fusion harnesses the powers of the Sun and provides an unlimited and cheap source of energy, while being pollution-free. Capturing the powers of the heavens has been fantasized in the past as mere science fiction, however this fiction may become a reality.

Although development is still in its infant stages, a new breakthrough by lead author Omar Hurricane, from the National Ignition Facility at the federally-funded Lawrence Livermore National Laboratory, published an article in Nature that announced researchers saw a net gain in energy following a fusion reaction. The reported results show almost 2 times more energy coming out of the reaction than what went into it. What does it take to run a reaction of this sort? One hundred and ninety-two of the world's most powerful lasers aimed at a 1 centimeter gold cylinder called a hohlarum. It is a small capsule that contains an extremely cold mixture of hydrogen isotopes. As the laser heats the capsule, the hydrogen is heated and subsequently compressed to 1/35 of its original size. Co-author Debbie Callahan described it as "compressing a basketball down to the size of a pea." If the compression is high and uniform enough, nuclear fusion will take place and in the nanoseconds that follows the capsule implosion, neutron energy is released.

Naturally this process occurs in the cores of stars. Unlike nuclear power, which relies on fission, the splitting of atoms, fusion occurs when atoms collide at a very high speed to form a new molecule. In this case, matter is not conserved and some mass dissipates as energy. When this process occurs in large quantities, a chain reaction of fusing releases more and more energy until there is self-sustaining energy source. This process called ignition is the ultimate goal of physicists. Though physicists have obtained net energy, there is still a considerable amount of work that lies ahead before the ignition is reached. Pulse shaping, which describes how lasers hit the hohlarum needs to be further developed, as the shape of implosion is not spherical which is desired for ignition. Additionally the 192 lasers used produce exponentially more energy than what was used in the reaction. It is likely that harnessing the full energy of these lasers may result in higher net gains.

Physicists are essentially trying to create a small, controlled star, so it is no wonder such a long and difficult process. However, the payoff is a limitless power supply. And the waste produced in these reactions are not radioactive, unlike those from nuclear fission, giving it the clear advantage for being a top energy choice. However there is no telling how long it will take to achieve ignition. The most optimistic scientists predict at least few decades.

O. A. Hurricane, D. A. Callahan, D. T. Casey, P. M. Celliers, C. Cerjan, E. L. Dewald, T. R. Dittrich, T. Döppner, D. E. Hinkel, L. F. Berzak Hopkins, J. L. Kline, S. Le Pape, T. Ma, A. G. MacPhee, J. L. Milovich, A. Pak, H.-S. Park, P. K. Patel, B. A. Remington, J. D. Salmonson, P. T. Springer & R. Tommasini. "Fuel Gain

Exceeding Unity in an Inertially Confined Fusion Implosion." Nature 506.7488
    (2014): 343-48.
http://www.nature.com/nature/journal/v506/n7488/full/nature13008.html
"NIF Fuel Gain Named a 'Top Breakthrough' of 2014." Science & Technology. Lawrence
    Livermore National Laboratory, https://lasers.llnl.gov/news/science-technology
RT News (http://on.rt.com/a3co1k)

The Key to Unlimited Energy: Nuclear Fusion and its Formidable Challenges

by Chad Redman

Modern nuclear power plants employ nuclear fission to produce energy with no emission of CO_2. Unfortunately, fission reactions create large amounts of radiation and have, consequently, awoken serious public concerns. Incidents similar to the recent reactor failure in Fukushima bring to light the dangers of harnessing the power of fission. Furthermore, current nuclear power plants produce significant quantities of hazardous waste; spent cores from these plants emit high levels of radiation and the long-term storage of such waste is an unresolved challenge.

Given steep resistance to nuclear fission technology, an alternative form of nuclear power presents itself – nuclear fusion. Fusion is a veritable holy grail of energy production. Nuclear fusion could satisfy human power needs for millions of years.

Current visions of fusion technology work, in simplest terms, by adding a neutron to a hydrogen atom, producing helium. During this process, a small quantity of mass is lost, and in accordance with the equation $E=mc^2$, energy is released. This same fundamental process is responsible for powering stars and hydrogen bombs, and boasts some serious engineering problems. Containing and capturing large-scale fusion energy is currently beyond our capabilities. While small-scale fusion has been demonstrated as possible, projects for experimental fusion power plants are only just developing. The foremost example of this is the International Thermonuclear Experimental Reactor (ITER) being built in France.

A multinational project, the ITER exemplifies just how difficult fusion reactions are to contain. The ITER will inject the fuel for the fusion reaction into a vacuum chamber, heated in excess of 100 million °F. This creates plasma, which will be contained in the vacuum chamber using a magnetic field. One of the major goals of the ITER will be to test the ability of a magnetic field to contain plasma long enough and at a high enough temperature for fusion to occur. Ideally, the ITER will prove that a fusion reaction can produce more energy than is required to initiate it, and that it can be safely contained.

Dresselhaus, M. S., Thomas, I. L. (2001). Alternative Energy Technologies. Nature, 414, 443-452.

"Provide Energy from Fusion." National Academy of Engineering. N.p., n.d. Web. 02 Mar. 2015.

Major US Nuclear Energy Provider Fights for Clean Energy Money in Order to Survive
by Trevor Smith

Exelon, the largest producer of nuclear power in the United States, is a major force behind an Illinois bill that would direct additional funds toward energy producers who do not create greenhouse gases. The funds would come from a small fee placed on power bills of consumers of these types of energy. Opponents of the bill, however, claim that the bill favors nuclear power over other forms of energy through arbitrary restrictions on who qualifies for the funds, and that since nuclear energy production leaves behind radioactive waste, it isn't truly a form of clean energy (O'Connor 2015).

The essence of the bill is that Illinois energy makers who do not produce greenhouse gases should be rewarded with a $2 credit for each household that consumes their energy (O'Connor 2015). The credit, which would simply be added onto consumers' power bills, is designed to help keep clean energy competitive with natural gas, the price of which has recently fallen to historically low levels. Exelon claims that because of the effect of natural gas on falling energy prices, it can no longer afford to keep three Illinois power plants in business; the bill serves as a sort of a bailout for nuclear energy (Henry 2015).

Opponents of the bill oppose it for two central reasons. First, various restrictions in the bill on factors such as the size of the energy plant would exclude much of Illinois's hydroelectric power, and other restrictions that would effectively shut down certain solar and wind plants from getting the credit are also in the draft of the bill currently proposed. Second, and perhaps more interestingly, opponents of the bill claim that nuclear energy shouldn't be classified along with energy, solar, and hydroelectric as clean energy. Although it doesn't produce greenhouse gases (Henry 2015), the bill's challengers note that nuclear energy still produces radioactive waste the storage of which causes a number of issues. Furthermore, nuclear energy's reliance on a finite supply of uranium-235 makes it a non-renewable source of energy, unlike the other forms of clean energy the bill aims to support.

The bill's introduction into the Illinois legislature is not just an attempt by a large nuclear company to survive, then, but a public debate on whether nuclear energy should be hailed as clean energy and embraced as part of the future of the United States' energy grid. This significant battle for nuclear energy occurs as a March 30 Gallup Poll notes that 51% of Americans support nuclear energy, although most Americans also prefer to prioritize renewable energies like solar and wind (Riffkin 2015).

Henry, Ray. "Nation's biggest nuclear firm makes a play for green money". Penn Energy, March 30, 2015.
http://www.pennenergy.com/articles/pennenergy/2015/03/nuclear-energy-exelon-makes-play-for-green-money.html

O'Connor, John. "Low-carbon bill would aid Exelon nuclear plants", Penn Energy, February 27, 2015.
http://www.pennenergy.com/articles/pennenergy/2015/02/low-carbon-bill-would-aid-exelon-nuclear-power-plants.html

Riffkin, Rebecca. "US Support for Nuclear Energy at 51%", Gallup, March 30, 2015.
http://www.gallup.com/poll/182180/support-nuclear-energy.aspx

VEHICLES

Vehicles are changing rapidly in response to concerns about photochemical smog, micro-particulates, and carbon dioxide emissions. They are also changing because partial or full electric drivetrains are potentially far more responsive than internal combustion engines. Gasoline/electric hybrids are now widely available and increasingly affordable. They will shortly appear in the most advanced luxury vehicles even though they began as a move toward more economic driving. These will be replaced in time by plug-in hybrids, and in the not-too-distant future, all-electric vehicles, exporting the pollution associated with vehicles elsewhere in the form of much more efficient and cleaner central power plants—many of them running on non-polluting renewable energy. In parallel, hydrogen fuel cells are increasingly used for larger commercial vehicles where the volume of the fuel supply is less important than in personal vehicles. There are many other developments in the works, some of which are discussed below.

Those Cool Electric Cars
by Hannah Brown

Previously, electric cars seemed to be on par with self-driving cars, and hover crafts. Unfathomable. But as the need to protect our habits while preserving the environment has increased, and studies continue to show us the inevitable degradation we have already caused, the push for affordable, electric cars has increased.

Powered by electricity stored within a battery pack in the car, electric cars do not need to burn the fuels that we are currently dependent on, that contribute greatly to the increase in greenhouse gases. Thus, electric cars are fundamentally environmentally useful. Now, in addition to their alternative fueling capacities, a study finds that they also assist in ameliorating the urban heat islands effect—a phenomenon of urban centers being hotter microclimates due to the activity in them and the concrete that covers them.

The urban heat island effect is also worsened by the heat that internal combustion engines—the type used in traditional cars—release. Researchers are now arguing that electric cars, with their cooler engines, could help fix this heat problem. Published in Nature, the study titled "Hidden Benefits of Electric Vehicles for Addressing Climate Change" from Michigan State University concluded that electric cars produce only one-fifth of the heat per mile generated by normal cars (Lee *et al.* 2015).

This study found that replacing conventional cars with electric ones would have a truly dramatic effect on the urban heat island effect. Using Beijing's blazing 2012 summer as a model, the researchers concluded that using electric cars would have reduced the general temperature by 0.94° C, which would in turn have reduced air conditioning usage by 14.44 kWh, which would also reduce the daily CO_2 emissions by 10,686 tonnes. The important step to recognize here is that in reducing the general temperature of the city, people will reduce their A/C usage, which is a significant contributor to climate change (CityMetric.com).

As you can see, the domino effect is huge, and good and healthy in this case. For these reasons, among others, the researchers argue that we should accelerate the process of replacing traditional cars with electric ones, despite the fact that they are currently more expensive, and environmentally dubious, to build.

Li, Canbig; Cao, Yijia, Cao; Zhang, Mi; Wang, JIanhui, Wang; Liu, Jianguo; Shi, Haiqing & Geng, Yinghui. Hidden Benefits of Electric Vehicles for Addressing Climate Change. Nature.com. March 19, 2015.
CityMetric Staff. 2015. Electric cars could help cool down our cities. CityMetric.com March 23, 2015

Hydrogen Powered Trams Make Their Way in China

by Hannah Brown

In Qingdao, a western city in China's Shandong region, a new kind of public transportation is getting its start. Developed by Sifang, a state-owned manufacturer, this hydrogen-powered tram, introduced in mid-March 2015, is making all kinds of noise.

While hydrogen technology has been used to power cars and even buses, Sifang's tram is the first time the technology is being applied to rail public transportation. As it is no secret that

China's infrastructure contributes significantly to the release of pollutants, this new technology is promising for environmental reasons. It is not only beneficial in alleviating dangerous emissions, its development is an admission by China that all countries must be concerned about climate change, and need to be making significant strides to combat it.

While it may look like a speedy bullet train, this new tram only travels about 70 kilometers per mile. However, it has the capacity to keep going for about 100 kilometers before needing to refuel. And, when the time comes to stock up on hydrogen, the refueling process takes only three minutes, speeding up transportation times significantly (YouTube.com).

This new tram is efficient, and its creators even say it will cause a reduction in operation costs as well. Additionally, using hydrogen fuel cells instead of traditional fuel means that the only biproduct from using the trams is water. In addition, because the fuel cells are kept under 100 °C, the trams do not produce any nitrogen oxides either. Obviously, this new tram is a cleaner solution to transportation (iflscience.com).

As the trams can carry about 380 passengers, there is promise for the greater population's transportation needs in this reformatted type of tram. There is just one tiny problem though: across China, there are currently only about 83 miles of tram tracks, established in only seven cities. No matter, Sifang isn't worried. They know that China is interested in the new tram technology and, in the next five years, plan on spending 200 billion yuan ($32 billion) developing tram tracks, multiplying the existing amount by tenfold. In addition to the new 1,200 miles of tracks, they will also buy more trams, traditional and hydrogen-powered.

While construction must happen to ensure the use of these hydrogen-trams, some cities are already adopting the new technology. For example Foshan, with a population of 8 million, is planning to start its tramline construction later this year. The city is pretty vested in in Sifang's success; they invested $72 million last year to make sure that Sifang manufacturing plants would occur in their vicinity. Together with Sifang they are also creating a national hydrogen-power research center. Providing jobs for their constituents, and clean transportation, Foshan becomes an interesting model of government–one that encourages clean energy, while recognizing the economic boon this industry can bring to its citizens (Bloomberg.com).

New China TV. World's first hydrogen-powered tram rolls off assembly line. March 20, 2015. [video]. Retrieved from https://www.youtube.com/watch?v=o9QAV_orYsc
Alford, Justine. "China Develops World's First Hydrogen-Powered Tram." IFLScience.com. March 24, 2015

Bloomberg. March 25, 2015. China's Hydrogen-Powered Future Starts in Trams, Not Cars (http://www.bloomberg.com/news/articles/2015-03-25/china-s-hydrogen-powered-future-starts-in-trams-not-cars)

Solar Impulse, the Solar Airplane, Journeys Around the World

by Nour Bundogji

Solar Impulse, a Swiss long-range experimental aircraft, is the only airplane able to fly on solar power. However, it took project leaders Andre Borschberg and Bertrand Piccard more than one attempt to reach this accomplishment.

The first Solar Impulse aircraft—registered as HB-SIA—was primarily designed as a demonstration aircraft. Lessons learned from this prototype were incorporated in Solar Impulse 2, the Round-The-World-Solar-Airplane. "This revolutionary single-seater aircraft made of carbon fiber has a 72 meter wingspan (larger than that of the Boeing 747-8I) for a weight of just 2,300 kg, equivalent to that of a car" (SolarImpulse.com). The Solar Impulse 2 also includes 17,000 solar cells built into the wing supply of four electric motors (17.5 CV each) with renewable energy. These solar cells recharge lithium batteries, weighing 633 kg (2077 lbs), allowing the aircraft to fly at night. The Solar Impulse 2 also has electrolytes, invented by Solvay, which help increase the energy density of the batteries. Furthermore, Bayer Material Science is allowing the use of its nanotechnologies and Décision is using carbon fibers that are lighter in weight than any previously seen.

With all this development and innovative technology going into the production of this solar airplane, its inaugural flight took place on 2 June 2014, piloted by Markus Scherdel. The aircraft averaged a ground speed of 30 knots, and reached an altitude of 5,500 feet. In its first intercontinental flight, the Solar Impulse 2 completed a 19-hour trip from Madrid, Spain to Rabat, Morocco. On this trip, Solar Impulse set two world records for manned solar-powered flight. The first was for the longest distance covered on a single trip—that of 1,468 km and the second was for a groundspeed of 117 knots (216 km/h; 135 mph).

The Solar Impulse took off for its fifth flight from Mandalay, Myanmar at 21:06 UTC, to Chongqing, China, reported Jonathan Amos, BBC Science Correspondent. Bertrand Piccard will fly the zero-fuel airplane on about 1375 km (742 NM) for an estimated time of 20 hours.

The project expects the circumnavigation of the globe to be completed in a total of 12 legs, with a return to Abu Dhabi in a few months' time. No solar-powered plane has ever flown around the world, and I am excited to watch the Solar Impulse 2 completes its groundbreaking journey.

Amos, J. Solar Impulse plane lands in China. BBCNews.com. March 30, 2015.
        http://www.bbc.com/news/science-environment-32110747
Solar Impulse. Solar Impulse RTW. March 30, 2015.
        http://info.solarimpulse.com/en/our-adventure/solar-impulse-2//#.VRmAPlxcT8k

Drone Technology Now Used to Monitor, Preserve, and Restore the Environment
by Jessie Capper

Scientists and environmentalists have discovered new uses for drones in the green sector—from observing wildlife and natural habitats to protecting areas with overharvested resources (Carroll Feb 20, 2015).

Chris Zappa at the Lamont Doherty Earth Observatory manages and operates drones that fly to areas in the Arctic. These drones monitor ice melt using infrared cameras to collect data on temperature changes and melt water coming off the ice (Carroll Feb 20, 2015). Normal oceanographic techniques are not capable of obtaining sufficient data; therefore, the drones have been extremely beneficial for the scientists at Lamont. Zappa also suggested that they will soon expand the drones' uses to "drop off micro buoys that will measure the temperature and salinity of the water" (Carroll Feb 20, 2015).

Similar to Lamont, the Woods Hole Oceanographic Institution is using an underwater drone to observe the base of the polar ice (Carroll Feb 20, 2015). Woods Hole has also used a similar done to monitor the deep-water behavior of great white sharks. Woods Hole scientists attached an electronic tag to a shark, and programmed the drone to follow it around underwater; as a result, these scientists were able to collect "stunning, and completely unexpected data, on how great whites hunt" (Carroll Feb 20, 2015). Underwater drones have improved scientists' understanding of shark behavior while flying drones are being used by environmentalists and scientists to protect whales and rhinos from being killed by hunters.

The Sea Shepherd Conservation Society sends flying drones to observe "pilot whales. If the whales get too close to the Faroe Islands, where locals have yearly hunts, small boats are sent out to chase the whales away" (Carroll Feb 20, 2015). According to Paul Watson co-found of The Sea Shepherd Conservation Society, 1,300 whales were killed from June to September; however, only 32 were killed last year under the Sea Shepherd's patrol (Carroll Feb 20, 2015).

Furthermore, Spanish engineers have developed a drone to fight rhino poachers in national parks in Africa. The drone, equipped with thermal vision, sound identification, and onboard processing data, reports any irregularities in real time (NBCNews Mar 23, 2015).

In addition to protecting wildlife, a company called BioCarbon Engineering has designed a drone to restore habitats by planting trees effectively and efficiently. Their ultimate goal is to rebuild global forests that have been destroyed due to lumbering, mining, agriculture, and urban expansion. Other scientists are amidst development of a prototype drone that can "shoot pods containing germinated seeds as well as nutrients and fertilizer to support the tree as it begins to grow"

(Carroll Feb 20, 2015). Lauren Fletcher—CEO and co-founder of BioCarbon Engineering—estimates that their technology could plant at least 36,000 trees a day, amounting to ten times the hand planting rate (Carroll Feb 20, 2015). The multiple uses of drones in the environmental sector have been extremely advantageous for monitoring and protecting wildlife; hopefully BioCarbon Engineering's prototype will expand these benefits to habitat restoration.

Carroll, Linda. "Eco-Drones Aid Researchers in Fight to Save the Environment." NBC News. February 20, 2015. Accessed March 31, 2015.
http://www.nbcnews.com/tech/innovation/eco-drones-aid-researchers-fight-save-environment-n309131.
"Ranger Drone Could Help Fight Against Rhino Poaching." NBC News. March 23, 2015. Accessed March 31, 2015.
http://www.nbcnews.com/science/environment/ranger-drone-could-help-fight-against-rhino-poaching-n328561

The Future of Car Racing: Electrifying the World of Motorsports

by Jessie Capper

People often associate motorsports with gas-powered racecars and loud noises; however, the future of motorsports is changing as companies like Formula E, Virgin, Mario Andretti, and Porsche are proving the viability of electric car racing (Einstein March 16, 2015). Formula E is a race series committed to the promotion of clean energy and sustainability through improving the image and perception of electric vehicles (FIA Formula E). The first Formula E race was held in Beijing last

September 13th, and is quickly spreading through cities around the world.

Virgin and the racing family dynasty of Mario Andretti were of the first few companies to support the development of these battery-powered racecars. Virgin, Andretti, and other racers in the Formula E series are excited by the potential the Formula E series presents to skeptics of battery cars. According to recent tests, a Formula E racecar is only a few tenths miles per hour slower than a classic, gasoline engine Formula 1 car (Einstein March 22, 2015). These electric cars can launch from 0 to 60 miles per hour in barely 2.8 seconds. Furthermore, due to their wheel-spinning torque, electric racecars can reach a top speed of roughly 140 mph during racing (Einstein March 22, 2015). According to Virgin owner Sir Richard Branson, "Twenty years from now, the smell of exhaust will be as rare (in cities) as the smell of cigarette smoke is in a restaurant today... Unless you have sports like this, you'll never get to a world that's carbon neutral by 2050" (Einstein March 22, 2015).

Unfortunately, recent Formula E races are also exposing a few of the downsides to electric propulsion. "Range anxiety" is one of the biggest disadvantages of regular battery cars. During the 1.4-mile Miami street course, Formula E racers were only able to make it halfway through the 39-lap run due to the lithium-ion batteries' excessive weight. These batteries make up nearly 40% of the 1,950-pound weight of the 1-seat, open-wheel Formula E racers, and caused 20 drivers to pull into their pits and continue to race in a second car (Einstein March 22, 2015). This issue does not seem to be of much concern to team manager Andretti, however, as he claims Formula E's technology is steadily improving. Formula E's target is an 8% annual increase in the amount of power the racecar batteries can store. As a result, Andretti and other companies partnered with Formula E hope they will eliminate pit stops entirely by the end of the decade—an accomplishment that traditional car racing is yet to achieve (Einstein March 22, 2015).

Eisenstein, Paul. "Formula E Racers Set Out to Change the World ... Quietly." NBC News. March 22, 2015. Accessed April 8, 2015.
    http://www.nbcnews.com/business/autos/formula-e-racers-set-out-change-world-quietly-n327071
Eisenstein, Paul. "Porsche May Add EV to Line-Up." The Detroit Bureau. March 16, 2015. Accessed April 8, 2015.
    http://www.thedetroitbureau.com/2015/03/porsche-may-add-ev-to-line-up/
FIA Formula E (http://www.fiaformulae.com/)

"Bio-bus" Runs on Treated Sewage and Food Waste

by Liza Farr

In November 2014, GENeco, a subsidiary of a UK utility company, debuted a new bus line that is powered by human and food waste (BBC, Nov 20, 2014). The so-called "poo bus" carries about 10,000 passengers each month between the Bristol airport and the Bath city center. The 40-seat bus can travel up to 186 miles on one tank, which can be produced from the annual waste of about five people. The combustion engine on the bus is similar in design to diesel equivalents. The gas is generated through anaerobic digestion, in which oxygen starved bacteria break down the biodegradable material to produce methane-rich biogas. The gas is processed over 12 to 18 days, where it also undergoes upgrading, carbon dioxide and impurities are removed, and propane is added. The result is a fuel with virtually odor-free emissions (Frangroul, Feb 3, 2015). By the end of processing, the composition of the biogas is the same as natural gas. The compressed gas is stored in dome-like tanks on the roof of the bio bus (BBC, Nov 20, 2014).

The bus produces up to 30% less carbon dioxide than conventional gasoline engines, and harmful emissions would be reduced by as much as 97% if fossil fuels were replaced with this human waste gas (Opelka, Feb 3, 2015). The bus was launched just as Bristol became the European Green Capital of 2015 (BBC, Nov 20, 2014). GENeco hopes the bus serves to educate people about fossil fuel use and using waste productively. However, as of yet there is little discussion of expansion of the bus line to include multiple bio buses. It is unclear whether the bus is economic, or simply a novel reputation piece to bring awareness to the issue of waste and energy. The Bristol sewage plant generates 17 million cubic meters of biomethane annually, which is enough to power 8,300 homes (BBC, Nov 20, 2014). GENeco does sell the biomethane to homes in the area, although it is unclear how many. The bio bus and GENeco's innovative use of human and food waste is still an exciting frontier in renewable energy and in closing the waste loop.

BBC. "UK's first 'poo bus' goes into service between Bristol and Bath." Nov 20, 2014. [http://www.bbc.com/news/uk-england-bristol-30115137]

Frangoul, Anmar. "Flush with success: The UK's first poop-powered bus." Feb 3, 2015. [http://www.cnbc.com/id/102375325#]

Opelka, Mike. "Yup, There's a 'Poop-Powered' Bus Out There...and It Works." Feb 3, 2015. [http://www.theblaze.com/stories/2015/02/03/yup-theres-a-poop-powered-bus-out-there-and-it-works/]

Green Cars are Being Replaced by SUVs: The Need for Cheap

by Liza Farr

The green car industry is interesting this year as gas prices plummet, but technology progresses. Examining the big picture, the US Energy Information Administration reported that as of 2011, over 11 million alternative fuel vehicles were being driven in America, up substantially from just 250,000 in 1995 (Kennedy, Apr 22, 2015). Of these, 2 million are hybrid gas-electric and diesel-electric, while the majority of the other alternative fuel cars are flexible fuel vehicles that can operate on gasoline or gasoline-ethanol blends (Kennedy, Apr 22, 2015). Although 11 million may seem substantial, this statistic must be put in perspective relative to the over 250 million cars and trucks that operate on gas or diesel (Kennedy, Apr 22, 2015). Overall, the average fuel economy of new vehicles has improved from 20.1 mpg in October 2007 to 25.4 in March of 2015, marking a 26% improvement over 8 years (Kennedy, Apr 22, 2015).

These statistics are interesting when placed in the context of the Kelley Blue Book's annual 10 best green cars list. The list includes well-known cars such as the Tesla Model S (#6), the Toyota Prius (#4), and the Nissan Leaf (#3). A senior associate editor at Kelley Blue Book stated that falling gas prices are not having as large of an effect on demand for green cars as might be expected because customers know that the fluctuating nature of gas prices means they will likely go back up, making green cars an economically sound option (Kennedy, Apr 22, 2015). Additionally, they note that car markers are expanding their customer market by introducing a wider diversity of cars, from the modestly priced Jetta TDI, which gets 46 mpg, to the luxury Tesla Model-S, which is entirely electric and has a 253 mile range (Kennedy, Apr 22, 2015, and Smith, Apr 23, 2015). Unfortunately, some statistics paint a less glowing picture of the green car industry. Twenty-two percent of people who traded in their hybrid or electric car exchanged it for a new SUV, up from 18.8% last year (Motavalli, Apr 22, 2015). Forty-five percent of 2015 trade-ins have been replaced with an alternative fuel vehicle, down from 60% in 2012 (Motavalli, Apr 22, 2015). At current gas prices of $2.27 on average, it would take 10.5 years to pay off the difference in price between a Camry and a Camry Hybrid (Motavalli, Apr 22, 2015). Analysts say if gas prices increase to $4 again, there may be another shift toward green cars, but even at that price it would take 5 years to pay off that difference (Motavalli, Apr 22, 2015).

Advocates for green cars are noticing this economic disadvantage of green cars as a true detriment to reducing automobile emissions. The Volkswagen e-gold was number two on Kelley Blue Book's top green car list, but only 181 cars were sold in March (Goreham, Apr 23, 2015). The BMW i3 sold under 1,000 in March, and only half of those were fully electric (Goreham, Apr 23, 2015). All of the electric vehicles on Kelley Blue Book's list sold 5,000 cars in total in March (Goreham, Apr 23, 2015). In contrast, the Toyota Prius, a hybrid car, sold 9,485 cars in its sedan form (Goreham, Apr 23, 2015). Although the Prius does use some gasoline, in contrast with electric vehicles that topped Kelley Blue Books list, since more have been sold, they can have a bigger impact in replacing dirtier cars. These numbers indicate that it is imperative for the green car industry to focus on reducing prices, even at the cost of incorporating hybrid vehicles rather than fully electric vehicles, in order to have a real impact on reducing automobile emissions.

Goreham, John. Toyota Prius outsells entire list of KBB top 10 green cars for 2015. April 23, 2015. [http://www.torquenews.com/1083/toyota-prius-outsells-entire-list-kbb-top-10-green-cars-2015]

Kennedy, Bruce. Green Cars Gain Momentum with American Drivers. April 22, 2015. [http://www.cbsnews.com/news/green-cars-gain-momentum-with-american-drivers/]

Motavalli, Jim. Americans are selling green cars and buying SUVs mother nature network. April 22, 2015. [http://www.mnn.com/green-tech/transportation/blogs/americans-are-selling-green-cars-and-buying-suvs]

Smith, Leo. The Spin: Tesla tops Auto Club's Green Car List. April 23, 2015. [http://www.ocregister.com/articles/car-659212-green-fuel.html]

Piston Power
by Alexander Flores

Achates Power has become actively engaged with nearly all of the leading commercial truck and passenger vehicle manufacturers worldwide in hopes to provide fundamentally better engines for the good of companies and the environment. Achates Power has become heavily recognized for its development and sale of the Achates Power Engine, an opposed-piston, two-stroke engine, which has proven to be much more efficient, clean, and durable for maximum vehicle performance. How exactly does it function? Essentially, in each cylinder, two pistons come together forming a combustion chamber then reciprocate outward. This motion by the pistons cyclically open and close the exhaust, which allows the intake of ports in order to manage the gas-exchange process. Their patented combustion system features optimized port and manifold designs, state-of-the-art fuel injector design and configuration, and a piston bowl shape that combines swirl with tumble motion allowing excellent mixing, air utilization, and charge motion for rapid diffusion flame propagation with minimal flame-wall interaction.

This results in clean and efficient combustion for the vehicle. The engine's three-cylinder configuration makes it optimal for gas-exchange since the exhaust events are timed to maintain turbocharger momentum without negative cross-charging effects. The engine also uses a dual crankshaft, Junkers Jumo-style arrangement, providing robustness, compactness, and mechanical simplicity. This compact design maximizes thermal efficiency due to a long stroke-to-bore ratio, which improves scavenging and reduces pumping losses.

Dynamometer testing of the Achates Power Engine has shown that it meets global emission standards, is fuel efficient, with 20+% versus the best diesels and 55+% versus the best gasoline engines, and less costly to manufacturer due to fewer parts. Recently, the company was just awarded a $14 Million Military Engine Project by the National Advanced Mobility Consortium in support of research and development work of the US Army Tank Automotive Research, Development, and Engineering Center (TARDEC). Their project is the Single Cylinder Advanced Combat Engine Technology Demonstrator, which will be a part of the Army's 30-year strategy to modernize tactical and combat vehicles. More power to the piston. Hoorah!

Achates Power 1 (http://www.achatespower.com/index.php)
Achates Power 2 (http://www.achatespower.com/opposed-piston-engine.php)
Achates Power 3 (http://www.achatespower.com/two-stroke-diesel-engine.php)
Achates Power 4
(http://www.achatespower.com/pdf/Achates%20Power_AdvancedCombatEngine_FINAL_33115.pdf)

Compressed Air Hybrid Vehicles?
by Emil Morhardt

The usual candidate power supplies for the non-fossil-fuel part of hybrid vehicles are chemical batteries, supercapacitors, and flywheels, all powered up using electricity, and generating electricity when their power can usefully replace or supplant the main power source, the internal combustion engine. But these types of electrical storage and the motor/generators they utilize are complex, sophisticated, and expensive, and have barely appeared at all in the developing parts of the world where fossil fuel use is growing fastest. Maybe there is a simpler, cheaper option. One possibility is compressed air energy storage. All you need is a tank (cheap), a reversible compressor (fairly cheap), and a way to link it to the engine. That last part is tricky because the general run of such systems work optimally at a specific pressure, with their performance falling off dramatically as pressures in the tank exceed or fall below optimum, which is what happens as the tank is being pressurized or depressurized. The simple solution, according to Brown *et al.* (2014), is to use inexpensive check valves on the tank to prevent over-compression and over-expansion, and an infinitely variable transmission between the compressor and the engine that can operate efficiently at the allowed range of tank pressures. The transmission adjusts by changing the number of thermodynamic cycles of the compressor executed per driveshaft rotation.

Nobody is doing this yet in road vehicles, the authors believe, because the standard approach uses the engine as a compressor and requires the addition of variable valve timing with expensive actuators, rather than the relatively inexpensive addition of a dedicated external compressor and transmission. Although the efficiency of the proof of concept system they cobbled together is not very high—only about 10% of the energy converted to compressed air comes back to drive the engine—the authors figure that if the exhaust heat of the engine were used to keep the air tank hot (which would cost nothing in terms of energy) a round trip efficiency of 47% might be realized at a cost lower than that of battery hybrids, and in a package that would last a long time and could be repaired using mechanical capabilities common in the developing world.

Brown, T., Atluri, V., Schmiedeler, J., 2014. A low-cost hybrid drivetrain concept based on compressed air energy storage. Applied Energy 134, 477-489.

Nissan's Electric Vehicle Fleet's Promising Future in Energy Storage and the Private Sector

by Melanie Paty

The electric vehicle market is booming; over the past year, sales of plug-in hybrids increased by 1067% and sales of pure electric models increased by 79%. Registrations in the United Kingdom alone quadrupled, largely due to the government's plug-in grant scheme, which offers grants of up to 35% of the full cost of electric cars and 20% for vans (Gov.UK 2015). Thus far, it has been far more successful with cars, reaching 25,000 grants, whereas only 1,117 grants have been awarded for vans.

These numbers are likely to rise following the mass-market vehicle-to-grid deal Nissan signed with Endesa, a European energy utility, in early March 2015. The scheme consists of an Endesa two-way charger and energy management system that can integrate power from solar panels and wind turbines, and allows electric vehicle drivers to charge their battery during low-demand, and use the electricity stored in the battery at home or sell it back to the grid during high demand. This is a realization of the electric car as energy storage model Governor Jerry Brown hopes to expand across California.

Nissan is the world leader in electric vehicle sales with 160,000 models sold around the world. It recently launched the Nissan e-NV200 in June 2014. This electric van costs $19,600 for lease and $24,250 for purchase, with operating costs of only $0.29 per mile. The average cost per mile of a standard minivan is $0.65 and $0.74 for the standard SUV (AAA 2014). The van can run up to 106 miles from eight hours of charging and shows live energy consumption and highlights nearby charging stations while in use. Running the car in "ECO mode" can increase the battery life by optimizing acceleration and climate controls and limiting the top speed to 60mph. However, using the AC system in either mode reduces range by approximately 20 miles. Also, the vehicle can feel shakey at high speeds unless there's cargo weighing down the back.

Nissan hopes to attract business and public sector fleets with the e-NV200. The five-seater version is being adopted as a zero emissions taxi in several cities including London, Barcelona, and New York. Nichols notes the surprising slow pace of the electric van market, as commercial van fleets are very cost-conscious and likely to see the benefit of all-electric vehicles.

Nichols, Will. Green Biz, "Can Nissan's all-electric van boost the appeal of commercial EV fleets?" April 13 2015 (http://www.greenbiz.com/article/nissans-all-electric-van-and-future-commercial-ev-fleets).

Nichols, Will. Green Biz, "EV income? Nissan on how electric car owners could sell extra power." March 7 2015 (http://www.greenbiz.com/article/could-your-ev-earn-you-cash-nissan-says-electric-car-drivers-could-sell-power-grid).

Gov.uk, "Plug-in car and van grants," (https://www.gov.uk/plug-in-car-van-grants/overview)

AAA, "Owning and Operating Your Vehicle Just got a Little Cheaper According to AAA's 2014 'Your Driving Costs' Study," May 9 2014, (http://newsroom.aaa.com/tag/driving-cost-per-mile/).

H2FC SUPERGEN: UK Hydrogen and Fuel Cell Research at its Finest
by Chad Redman

High on the list of fuels with the potential to reduce humankind's carbon footprint sits an abundant resource, hydrogen. Given such potential, the Hydrogen and Fuel Cell (H2FC) SUPERGEN Hub in the United Kingdom focuses on research and development for the hydrogen and fuel cell industry. The Hub is a multifaceted organization launched in May 2012 that combines scientific, economic, and policy-related R&D under one roof. Funding from the UK's Engineering and Physical Sciences Research Council draws in professionals from all three camps and encourages collaboration with the goal of advancing hydrogen and fuel cell use across the globe.

The H2FC SUPERGEN Hub is compartmentalized into several different divisions, each with its own specialization in the hydrogen and fuel cell industry. For example, one division of the Hub is the Systems group, focusing on integrating separate hydrogen and fuel cell units, as well as large system optimization. The research conducted by the Systems group includes topics such as developing a refueling supply chain.

At the other end of the Hub's research spectrum is the Socioeconomic research group. This group has published research on the socioeconomic feasibility of technologies such as hydrogen fuel cell cars, covering the economic and political aspects of the hydrogen and fuel cell industry. Considerations such as infrastructure cost and policy challenges are handled by the socioeconomic research group.

As an organization, the UK's Hydrogen and Fuel Cell SUPERGEN Hub is a sign that hydrogen may become a common source of energy in the not-so-distant future. Collaboration and funding for projects such as these give high hopes to those seeking more responsible ways to power our lives.

Chloe Stockford, Nigel Brandon, and John Irvine. "H2FG SUPERGEN: An Overview of the Hydrogen and Fuel Cell Research Across the UK." International Journal of Hydrogen Energy 40 (2015): 5534-543. Web. 30 Mar. 2015.

The Economic Benefit of Strict Auto Emission Regulations
by Chad Redman

The auto industry in the United States has been subject to increasingly stringent emissions regulations over the course of the past several years. This trend has reduced carbon and other emissions from vehicles dramatically, helping to cut air pollution significantly. However, at what economic cost has this progress been made?

CNBC's Margo Oge, former EPA director, argues that auto emissions regulations have done nothing but stimulate economic growth. Oge writes that the year 2014 saw the highest sales in decades, with American consumers buying up a staggering 16.5 million new vehicles. All the while, this updated and expanded fleet of vehicles did not accelerate greenhouse gas emissions beyond 2013 levels. This feat was accomplished through innovation by auto manufacturers, namely increased fuel economy.

In order to reach the engineering requirements of emissions control regulations, manufacturers have added significant investment to R&D programs which will allow them to stay on track with legislation. This investment manifests itself in jobs, pumping money into the pockets of US citizens and contributing to a growing economy. Oge writes that estimates predict investment due to greenhouse gas regulation will create 150,000 new jobs by the year 2020. In short, federal regulation over emissions may not only be good for our environment, but for our wallet too.

Oge, Margo T. "Why Auto-emission Rules Are Good for the Economy." CNBC. CNBC, 21 Apr. 2015. Web. 21 Apr. 2015.

China to use Electric Vehicles to Solve Energy Variability Problem
by Ali Siddiqui

Wang Zhongying, director of the China National Renewable Energy Center, has stated, "The electricity vehicle could be the big contribution for power system's stability and reliability." He has, also, contributed towards a study, which relies on Vehicle-to-Grid technology, titled the "China 2050 High Renewable Energy Penetration Scenario and Roadmap Study." This study's aim is to move China's energy away from the coal sector, have 85% of electricity come from renewable energy, and have greenhouse gas emissions be cut by 60%.

The aim of the plan is to have owners' of electric cars charge their car's batteries during the late afternoon. Then, when supply of energy on the grid is low and demand increases, the owners' can then sell it back to the grid. The reason behind having these interactions take place is because of the heavy reliance of Chinese energy on solar and wind power. This alternative energy source is variable because it depends on the sun shining or the wind blowing. The owner's stored electricity would be able to counteract that variable cost associated with alternative energy and still allow for electricity to be available in the grid during times of low energy.

The best electric cars have a range of 250 km and can store up to 40 kWh of electricity. In China, such an electric car costs around $40,000, and may be cheaper soon. China may subsidize the cost of the car in order to increase national demand.

Li Junfeng, director general of China's National Center for Climate Change Strategy and International Co-operation, however, has stated that China's top priority is to be among the developed countries by 2050 with respect to people's standard of living, which makes China's first priority economic development and not alternative energy. This makes Wang's plan visionary according to Li.

McMahon, Jeff. China: Electric Vehicle-to-Grid Technology Could Solve Renewable Energy Storage Problem. April 21st 2015.
http://www.forbes.com/sites/jeffmcmahon/2015/04/21/china-electric-vehicle-to-grid-tech-could-solve-renewable-energy-storage-problem/

BIOFUELS AND SYNFUELS

Whether or not it was a useful exercise to mandate corn ethanol into the fuel supply, it is clear that utilizing waste plant matter and other waste biological material from wastewater treatment plants and landfills as fuels is a good idea. It also seems likely that purpose grown crops that do not displace the food supply or natural ecosystems might be a good source of fuel; algae grown in sewage effluent or seawater for example. There are many other ideas in the wind as well.

Solazyme Algal Biofuel Production in the United States
by Mariah Valerie Barber

In early February 2014, in Galva, Iowa at the American Natural Products (ANP) facility and in Clinton, Iowa, at the Archer Daniels Midland Company (ADM), Solazyme, Inc., began its commercial production of algal biofuel and oil. Solazyme, a San Francisco based firm utilizes microalgae, which it refers to as "the world's original oil producer," in order to produce biofuel (Solazyme.com). Solazyme creates oil from microalgae by a process of industrial fermentation, during which the microalgae is not fed with solar energy, but with sugar, which results in the production oil. Using industrial fermentation speeds up the natural chemical processes, which algae undergo. Once the microalgae produce the oil, the oil is extracted and made ready for commercial use. Even before the facilities in Iowa opened, Solazyme has had facilities in both Peoria and Orindiúva, Brazil. Peoria has the capacity to manufacture 2,000 metric tons of oil per year whereas the new facilities now are each able to produce 100,000 metric tons of oil per year (Clean Technica.com). Solayzme, which claims to be the first oil producer, has potential to drastically transform the oil industry and its reliance on fossil fuels.

Despite the fact that Solazyme has potential to eliminate the reliance on fossil fuels and all of the negative environmental impacts that are associated with their extraction and production, the stocks and investment in the company have been steadily declining for the last year (Ashburn Daily.com). Since opening its commercial facilities in the beginning of 2014, Solazyme has not yet been producing at its capacity of 100,000 metric tons of oil per year (Solazyme.com). Additionally, because fossil fuel oil prices are extremely low, demand for Solazyme may have gone down. Currently Solazyme is not making much profit but has a substantial amount of operating capital. Many people believe that the profits of Solazyme and the prominence of its commercial use with take several years to become visible.

Solazyme, Inc. (http://solazyme.com/innovation/?lang=en)

Clean Technica (http://cleantechnica.com/2014/02/03/solazyme-begins-commercial-production-algal-oil-us/)

Ashburn Daily (http://www.ashburndaily.com/shares-of-solazyme-inc-sinks-by-4-33-for-the-week/312135/).

The Zero-Waste Zoo: Using Animal Waste as Fuel

by Alex Elder

On April 17th 2015, the Detroit Zoo launched a campaign on the crowdfunding site Patronicity to raise funds for an anaerobic biodigester. This piece of technology will be able to convert animal waste produced at the zoo into compost and methane-rich gas. The compost generated will be used to enhance the animal habitats, landscapes, and other public spaces at the zoo. The gas will be used to generate renewable heat and power for the 18,000-square-foot Ruth Roby Glancy Animal Health Complex, which is a state-of-the-art veterinary hospital located at the Detroit Zoo.

With more than 400 tons of manure generated annually at the Detroit Zoo, there is an abundance of this material, with majority of this product simply being disposed of. The anaerobic biodigester works by putting biomass inside a sealed tank, where naturally occurring micro-organisms digest the biomass. This process releases a methane-rich gas (also known as biogas) that can be used to generate energy. Because of the renewable energy system, the biodigester would help to reduce fossil fuel consumption and lower greenhouse gas emissions. Any leftover biomass material is rich in nutrients and can be used as fertilizer throughout the zoo.

The Detroit Zoo is currently seeking $55,000 from donors through their crowdsourcing campaign to establish this biodigester technology on their premises. If the goal is met, The

Michigan Economic Development Corporation will match the donation with an additional $55,000. Although the initial investment for the project is expensive, the biodigester will save the zoo nearly $80,000 in energy costs and another $40,000 in waste disposal fees. The Detroit Zoo has set a goal of becoming zero-waste by 2020.

Welch, S. (2015). With new machine, Detroit Zoo aims to turn waste into power. Crain's Detroit. http://www.crainsdetroit.com/article/20150419/NEWS/304199984/with-new-machine-detroit-zoo-aims-to-turn-waste-into-power

Patronicity. Detriot Zoo Biodigester – Green Spaces and Green Energy. https://www.patronicity.com/project/detroit_zoo_biodigester_green_spaces_and_green_energy

Solazyme Seeks Solutions
by Alexander Flores

Solazyme, an up and coming company from south San Francisco is pursuing a vision to improve our planet and human life by producing sustainable, high-performance oils and ingredients derived from the microalgae that accumulates in countless places. Solazyme is attacking some of today's biggest issues such as sustainability, resource scarcity, and resource traceability by identifying oil profiles that can improve both performance and longevity of products. These oil profiles are then produced with microalgae, providing alternatives to petroleum, vegetable oils, and animal fats. Solazyme's products may be useful in personal care, industrial usage, food, and fuel industries. Their main oil-producing microalgae strain was originally discovered over a century ago on the sap of a chestnut tree in Germany and was carefully selected after approximately over ten thousand screenings of other strains. This microalgae can grow without any light and can natively convert sugars directly into oils and other whole algal products in closed fermentation tanks. Typically, a microalgae strain is selected to produce a specific oil, which then goes through a drying or standard oil extraction stage. Some of these products include AlgaVia, Whole Algal Flour, Whole Algal Protein food ingredients, and alguronic acid. They claim that their high stability, high oleic oil is one of the healthiest oils in existence with low levels of polyunsaturated fatty acids, zero transfats, high omega 9 fatty acids, and low saturated fats. These uniquely derived oils may find themselves in products in your home in the near future, from a bar of soap to paint to tires to cooking oil and even cosmetics. Along with these particular household applications, these microalgae can also be used to develop biofuels that burn cleaner and perform better than petroleum-based fuels. For instance, Soladiesel BD is 100% algae-derived and can already be used with standard diesel engines without modification. This fuel is also fully compliant with the ASTM D 6751 specifications for Fatty Acid Methyl-Esther (FAME) based fuel that meets ASTM D 975, and significantly outperforms ultra-low sulfur diesel in total THC, carbon monoxide and particulate matter tailpipe emissions. Solajet, on the other hand, is the world's first microbially-derived jet fuel to meet key industry specifications for commercial aviation, ASTM D 1655. It is fully compatible with the existing infrastructure and offers a faster, farther, and greater payload by reducing wing heat stress, lowering flammability, lowering smoke emissions, increasing storage life and lowering maintenance costs overall. Solazyme seeks to decouple the production of oil from geography and reduce the ecosystem damage that unsustainable oils have caused. Essentially, these algal oils can replace fossil fuels at one end of the spectrum and unsustainable plant-based oils at the other.

Solazyme 1 (http://solazyme.com/innovation/?lang=en)
Solazyme 2 (http://solazyme.com/solutions/food/?lang=en)
Solazyme 3 (http://solazyme.com/solutions/fuel/?lang=en)
Solazyme 4 (http://solazyme.com/solutions/industrial/?lang=en)
Solazyme 5 (http://solazyme.com/solutions/personal-care/?lang=en)
Solazyme 6 (http://solazyme.com/company/?lang=en)
TAGS: Braemar Energy Ventures, Energy Ventures, Harris & Harris Group, Lightspeed Venture Partners, Roda Group, VantagePoint Venture Partners, Solazyme, Soladiesel, Solajet, Jonathan Wolfson, Peter Licari

Algae from Wastewater for Biofuels
by Dylan Goodman

A Recent study in the bioscience department at has been looking at the application of sewer water as feedstock for producing biofuels. The study took place over a five year period at a local water treatment plant in Houston. Researchers discovered they could grow high-value algae while simultaneously removing 90% of nitrates and upwards of 50% of phosphorus from the wastewater.

In recent years the algae industry has moved towards more high-value products for pharmaceutical and nutritional companies. The production of algae, however, is largely dependent on chemical fertilizers. (Bhattacharjee, Rice University, 2015) Unfortunately, chemical based fertilizers also pollute US waterways. Pollution from the two primary components of chemical fertilizers, nitrogen and phosphors, is one of America's most widespread and costly environmental challenges (EPA). There is currently a need to look towards other means of producing high-value biofuels. According Meenakshi Bhattacharjee of Rice University, the use of chemical fertilizers both reduces profits and puts the algae industry in competition with agricultural companies.

Current wastewater treatment facilities don't have an effective means of reducing nitrate and phosphorous levels from treated water, so using wastewater to produce algae has the potential to solve two problems at once. "The idea has been on the books for quite a while, but there are questions, including whether it can be done in open tanks and whether it will be adaptable for monoculture — a preferred process where producers grow one algal strain that's optimized to yield particular products" (Siemann, Rice University) The Houston Department of Public Works and Engineering helped set up twelve open-topped 600 gallon tanks. They tested various algae combinations in different tanks and all twelve of them showed significant growth. Compared to a similar study conducting in Kansas, the Rice University study proved four times as effective, however researchers believe this could be because the tanks in Houston were about 30 degrees warmer than the tanks in Kansas. Further studies may be better able to answer this question. Regardless, using wastewater to produce biofuels has the potential to reduce pollutants while simultaneously making algaculture more sustainable.

Press Release. "Rice University Study: Algae from Wastewater for Biofuels." Solar Thermal Magazine. N.p., 04 Apr. 2015. Web.
Bhattacharjee, Meenakshi, and Evan Siemann. "Low Algal Diversity Systems Are a Promising Method for Biodiesel Production in Wastewater Fed Open Reactors." Algae 30.1 (2015): 67-79.

Our Future in Feces: Vehicles Fueled by Biomethane

by Briton Lee

The British company Geneco has begun implementing prominent buses with a cartoon graphic detailing its power source: human waste. The concept of using waste to produce fuel is not novel, and is termed biomethane or renewable natural gas (RNG). It is most commonly used to power vehicles. The biomethane is collected from sewage treatment plants that process human waste while producing methane and carbon dioxide as byproducts. Typically, the resulting gases are simply released into the atmosphere, with plants in Oslo, Norway producing and releasing approximately 17,000 tons of carbon dioxide a year (Demerjian 2009). The use of RNG in buses allows the reuse of these organic molecules and extracts remaining energy from these sources before releasing carbon dioxide back into the atmosphere. Our carbon footprint is not only mitigated at wastewater plants, as the engines needed to run on RNG are more efficient than traditional diesel engines, eking more mileage for the same amount of carbon emissions (Chappell 2014). In fact, the Bio-Bus' engine allows it to cover 186 miles on a full tank, which is equivalent to the annual waste of five people. There are further avenues for pursuing sustainability with the production of RNG, which can be obtained from decomposing food waste. Food waste and food not suitable for human consumption can be recycled via anaerobic digestion into an alternative energy source, instead of being wasted in landfills and incinerators that squander the untapped methane. The methane can also come from other sources of feces, such as livestock. RNG is usually locally sourced, and there is potential to turn rural communities into alternative fuel producers. In California, proponents of biomethane claim that collecting the methane from the state's 1.7 million cows would potentially produce 8 million pounds of methane a year, rivaling approximately 150 gallons of gasoline (Demerjian 2009). Regardless, there are multiple sources available for biomethane, and implementing RNG is quite feasible with 60% of Sweden's natural gas vehicles running on RNG (US Dept. of Energy).

Chappell, Bill. 2014. "Poo Power: New British Bus Runs On Human Waste." NPR. (http://www.npr.org/blogs/thetwo-way/2014/11/21/365761662/poo-power-new-british-bus-runs-on-human-waste)
Demerjian, Dave. 2009. "Norway or the Highway: Poo Powers Oslo Buses." Wired. (http://www.wired.com/2009/01/oslos-buses-to/)
US Dept. of Energy. "Renewable Natural Gas (Biomethane)." (http://www.afdc.energy.gov/fuels/emerging_biogas.html)

Algae Produce More Biofuel When Starved of Nitrogen, But Why?

by Emil Morhardt

Algae, like all organisms, require nitrogen to produce amino acids, the building blocks of proteins, and necessary for DNA synthesis. When deprived of nitrogen, some species, such as the micro alga *Chlamydomonas reinhardtii* studied by Valledor *et al.* (2014), produce more lipids (oil) than normal, presumably as a stored energy source to tide them over until nitrogen again becomes available. These lipids could become the major source of biofuel if their production can be sufficiently ramped up. Valledor *et al.* wanted a better understanding of what was going on at the molecular level in the nitrogen-deprived algae so that they could eventually modify the species genetically to enhance oil production. They limited nitrogen, and quantified the changes in the cellular mix of protein and metabolic products (the proteome and metabolome), looking at the levels of over 1,200 proteins, 845 of which were recognized as enzymes mediating 157 known cellular metabolic pathways, half of those known for this species. Then they reintroduced nitrogen and followed the process further.

Under nitrogen limitation, there was a decrease in proteins involved in photosynthesis, but even though there was an increase in lipid production, there wasn't much change in the enzymes responsible for it. Changes in the levels of a good many other cellular molecules were also documented though, including a variety that look as though if their production were increased, lipid production might also increase. Thus, the authors uncovered a number of candidate proteins to genetically manipulate. The pathways are extremely complex, however, and it is clear that the nitrogen starvation leading to increased lipid production very substantially alters many aspects of the cellular metabolism. The next step is probably to increase the production of some of these enzymes and see what happens.

Results like this are the usual story in biology. The more closely one looks, the more complicated things get, and sophisticated tools like the high throughput mass spectrometry used here are necessary to get to the bottom of it. Increasing lipid production through genetic engineering will not likely be easy, but probably would not come about at all without this type of basic research. Until you have a fairly good understanding of a metabolic system, it is extremely hard to effectively manipulate it.

Valledor, L., Furuhashi, T., Wienkoop, S., Weckwerth, W., 2014. System-level network analysis of nitrogen starvation and recovery in Chlamydomonas reinhardtii reveals potential new targets for increased lipid accumulation. Biotechnology for Biofuels 7, 171. http://www.biotechnologyforbiofuels.com/content/pdf/s13068-014-0171-1.pdf

Biomass to Butanol via Engineered Yeast
by Emil Morhardt

Butanol is a four-carbon alcohol, next in size after 1-C methanol (wood alcohol), 2-C ethanol (drinking alcohol), and 3-C propanol (rubbing alcohol), so it shouldn't come as any surprise that yeast ought to be able to synthesize it out of sugar. And it burns like the other alcohols mentioned, so it is potentially a usable liquid fuel that could be mixed with gasoline (like ethanol, to increase it's non-fossil-fuel content), processed into other types of fuel, or used as commercial feedstock to make bio-based commercial plastics such as the PET (polyethylene terephthalate) used to make beverage bottles. Gevo, Inc., a company based in Englewood, Colorado but with it's only [troubled] production facility in Luverne, Minnesota, seems to be gradually overcoming myriad difficulties in commercializing biomass-based isobutanol, and is beginning to license its proprietary genetically-modified yeast, which produce more isobutanol than conventional ethanol-producing commercial varieties. Gevo hopes that these yeast will feel right at home in existing ethanol-production facilities (such as the Luverne plant, where they didn't do so well initially), and that all Gevo will have to do to get isobutanol out is to bolt on a module that separates the isobutanol from the water in which the yeast are living.

According to Brett Lund, the chief licensing officer at Gevo (as noted in the article cited below in Biomass Magazine) the clever way they do this is to create a vacuum above the yeast/water/butanol broth, which volatilizes the butanol which is then condensed out. This is similar to the distillation concept used with ethanol, but doesn't require boiling the broth which is expensive and kills the yeast.

Gevo's yeast engineering appears to allow them to produce yeast specialized to ferment optimally whatever kind of sugars are available (for example they have just signed a licensing agreement with Highlands EnviroFuels (chemengonline.com) which already has a syrup mill processing locally grown sugar cane and sweet sorghum, onto the back end of which they propose to "bolt" an isobutanol plant. Isobutanol is not as soluable in water as ethanol so it is considered a good potential marine fuel, apparently satisfying the requirements of the Coast Guard, and has been demonstrated to be convertible into a fuel suitable for powering USAF fighter jets.

http://www.gevo.com/about/company-overview/
http://biomassmagazine.com/articles/9607/isobutanol-to-the-rescue/

Study Shows Flawed Experiments used to Support Policies for "Low-Carbon" Biofuels

by Niti Nagar

According to John DeCicco, researcher at University of Michigan's Energy Institute, nearly all of the studies used to promote biofuels as climate-friendly alternatives to petroleum fuels are flawed and need to be redone. After reviewing more than 100 papers published over the span of more than two decades, DeCicco claims erroneous methodology has led to the false assumption that biofuels will limit emissions of carbon dioxide. Existing studies fail to correctly account for the carbon dioxide absorbed from the atmosphere when corn, soybeans and sugarcane are grown to make biofuels said DeCicco. He explains, "Almost all of the fields used to produce biofuels were already being used to produce crops for food, so there is no significant increase in the amount of carbon dioxide being removed from the atmosphere. Therefore, there's no climate benefit."

In his paper, DeCicco focuses on the carbon footprint model to evaluate the impacts of petroleum-based fuels and plant-based biofuels. Computing the total carbon footprint as a way to evaluate the total emissions of carbon dioxide and other greenhouse gases associated with the production and use of transportation fuels was a type of analysis introduced in the late 1980s. Since then, the results of many fuel-related carbon footprint analyses have led to widespread disagreement. Yet, despite these controversial methods, they were still advocated by environmental groups and were subsequently used by Congress as part of the 2007 federal energy bill's provisions to promote biofuels, which resulted in the US Renewable Fuel Standard and eventually California's Low-Carbon Fuel Standard.

In his analysis, DeCicco shows that existing carbon footprint analyses fail to properly reflect the dynamics of the carbon cycle by miscounting carbon dioxide uptake during plant growth. He emphasized that this process occurs on all productive lands, whether or not the land is harvested for biofuel, and concludes these modeling errors help explain the controversial results. Disagreements have been especially apparent when comparing biofuels, such as ethanol and biodiesel, to conventional petroleum-derived fuels such as gasoline and diesel.

DeCicco believes research should be focused on ways of removing carbon dioxide at faster rates and larger scales to increase net carbon dioxide uptake and effectively counterbalance emissions from the combustion of gasoline and other liquid fuels.

John M. DeCicco. The liquid carbon challenge: evolving views on transportation fuels and climate. Wiley Interdisciplinary Reviews: Energy and Environment, 2015; 4 (1): 98 DOI: 10.1002/wene.133

University of Michigan. "Closer look at flawed studies behind policies used to promote 'low-carbon' biofuels." ScienceDaily. ScienceDaily, 5 February 2015. <www.sciencedaily.com/releases/2015/02/150205122737.htm>.

How to Make Profits from Garbage
by Niti Nagar

A new patent, invented at the University of Texas at El Paso, can offer a lucrative alternative use to the rotting waste in landfill sites. Decomposing trash produces a landfill greenhouse gas, methane. Methane is also a fuel that can be used to produce electricity or heat. However most landfills do not produce enough methane to invest in collecting it for energy production. Alternatively, many landfills instead choose to burn the methane away, so the methane is at least reduced to the less the harmful, carbon dioxide. However chemistry professor Russell Chianelli, Ph.D., invented a patented process that shows how methane production can be profitable. He says, "We're wasting valuable methane by flaring it off," Chianelli said. "This process can help landfills make plenty of electricity to turn a profit by selling it back to the electric company."

The process consists of capturing and recycling the exhaust gas that is produced when landfill methane is burned to generate electricity. Recycling the exhaust gas into the landfill provides heat and additional moisture. Carbon dioxide, which is also found in the exhaust gas, will release additional methane once recycled within the landfill. Furthermore, Chianelli suggests that a portion of exhaust gas can be used to cultivate algae. He explains, "What makes the methane in landfills are the organisms that are feeding on decomposing waste. So what we need to do is feed them even more for more methane production." A portion of the cultured algae can be recycled to the landfill to increase methane output, while the other portion can be used to create biodiesels fuels. Chianelli has created a clean "zero-discharge system" that could change landfills across America.

Unvieristy of Texas at El Paso. "How to make a profit from rotting garbage." ScienceDaily. 31 March 2015. <www.sciencedaily.com/releases/2015/03/150331100859.htm>

One Step Closer to Renewable Propane
by Niti Nagar

A study published in Biotechnology for Biofuels, shows that researchers at the University of Manchester's Institute of Biotechnology (MIB), along with colleagues at Imperial College and University of Turku, have created a synthetic pathway for biosynthesis of the gas propane. This significant breakthrough is bringing the team a little closer to commercially producing renewable propane. Because natural metabolic pathways for the renewable biosynthesis of propane do not exist, scientists have developed an alternative microbial biosynthetic pathway. The team, led by Nigel Scrutton and Dr. Patrik Jones from Imperial College, modified existing fermentative butanol pathways using an engineered enzyme variant to redirect the microbial pathway to produce propane instead of butanol. The new pathway utilizes enzyme CoA intermediates that are derived from clostridial-like fermentative butanol pathways.

Director of the MIB, Nigel Scrutton says, "'The chemical industry is undergoing a major transformation as a consequence of unstable energy costs, limited natural resources and climate change. Efforts to find cleaner, more sustainable forms of energy as well as using biotechnology techniques to produce synthetic chemicals are currently being developed at The University of Manchester." This study provides new insight and understanding of the development of next-generation biofuels, a vital area of development as fossil fuels continue to dwindle.

Propane seems to be a good option for an energy source as its physicochemical properties allow it to be stored and transported in a compressed liquid form. Under ambient conditions it is a clean-burning gas, with existing global markets and infrastructure for storage, distribution, and utilization in a wide range of applications ranging from heating to transport fuel. For these reasons, propane is an attractive target product in research aimed at developing new renewable alternatives to complement currently used petroleum-derived fuels. Professor Scrutton comments, "This study focused on the construction and evaluation of alternative microbial biosynthetic pathways for the production of renewable propane. It also expands the metabolic toolbox for renewable propane production, providing new insight and understanding of the development of next-generation biofuels which one day could lead to commercial production."

Manschester University. "Scientists a step closer to developing renewable propane." ScienceDaily. ScienceDaily. 10 April 2015. <www.sciencedaily.com/releases/2015/150410083512.htm>.

Harnessing Energy from Household Plants
by Niti Nagar

Plants use sunlight to create energy through the process of photosynthesis. Photosynthesis has fueled plant life for about 450 million years, but a research team at the University of Georgia headed by Ramaraja Ramasamy has found a way to use this process to benefit growing human demands for energy. By interrupting photosynthesis, the team extracted the energy generated from plants to create truly green energy.

During photosynthesis, plants consume carbon dioxide and water to create energy in the form of starch and sugar. Water molecules are split by enzymes and oxygen is produced in the process. Hydrogen ions and electrons are also formed and released at this step. The team used nanotubes to create a siphon that collects the newly freed electrons before they can enter the electron transport chain. The nanotubes are placed in the chloroplast. By alternating thylakoids, structures inside chloroplast where light-dependent photosynthesis occurs, the electrons are then directed down a wire to generate an electrical current.

When the current of the plant was tested against one from a similar sized solar cell, the current generated from the plant was about twice as strong as the one in the cell. These results are surprising because plants are much less efficient at generating energy from the sun than solar cells. Solar panels have proven to be up to ten times more efficient than plants at generating electricity. The team published their results in the April 2015 edition of Energy & Environmental Science.

Although these results are encouraging, this technology is still in its infancy and there are no practical uses for it. However, theoretically it could be used to create energy in the household that could power everyday items like lights, TVs, and computers. If developed on a large scale, it could be used to power entire grids, but it would have to demonstrate efficiency and cost-feasibility. However, using plants in this manner would likely boost the amount of foliage planted, which will have the added bonus of better scenery and decreased air pollution.

University of Georgia. "Houseplants could one day power TVs, computers, and more." IFLScience. 7 January 2015. <http://www.iflscience.com/technology/houseplants-could-one-day-power-tvs-computers-and-more>.

Calkins, Jessica O., Yogeswaran Umasankar, Hugh O'neill, and Ramaraja P. Ramasamy. "High Photo-electrochemical Activity of Thylakoid–carbon Nanotube Composites for Photosynthetic Energy Conversion." Energy & Environmental Science 6.6 (2013): 1891-900. 22 Apr. 2015.

The Success of Incentivizing Renewable Energy Projects from Landfills
by Shannon O'Neill

Every year, a total of 164 million tons of waste is disposed of in landfills. This has created a concern for waste management, particularly due to the fact that landfills are the third largest source of methane, a greenhouse gas that negatively effects the environment. However, methane has been developed as an energy source, in which it is recaptured and used to power homes and businesses. Today, there are more than 630 landfill gas energy projects that together, produce 16.5 billion kilowatt hours of electricity a year. This is enough energy to provide for 1.5 million homes.

The EPA has recognized 450 landfills sites capable of developing such energy projects, however there are heavy prices associated with such projects, limiting the benefits of removing methane while simultaneously creating a renewable energy source. Because of the great environmental and energy advantages, the government has used various tools in order to incentivize such projects.

The researchers looked at four different policy tools (a renewable portfolio standard, production tax credits, investment tax credits, and state grants) in order to conclude which policies incentivize projects the most successfully. It was concluded that only the renewable portfolio standard and investment tax credits significantly and positively contributed to the success of renewable energy projects. The renewable portfolio standard allows landfills to gain a profit from both the energy and the renewable energy credits it sells. The investment tax credits are given in order to install the necessary renewable energy facility at a subsidized price. These two policies alone provide a net benefit of 41.8 million by reducing methane and creating a renewable energy resource.

This provides powerful insight when looking to the future, specifically looking at the 450 landfill sites able to develop renewable energy projects. There should be greater emphasis on the two aforementioned policies as the benefits are quite large not only in profits, but also in energy and the environment, in order to create the most renewable energy possible while concurrently reducing greenhouse gas emissions of methane.

Li, S., Yo H.K., Shih, J.S., (2015). Renewable Energy from Landfills. Resources for the Future, 188:18-19.

Resources for the Future (http://www.rff.org/RFF/Documents/RFF-Resources-188_Featurette-LiYooShih.pdf).

The Importance of Wood as a Renewable Energy Resource
by Shannon O'Neill

The importance of wood as a renewable energy resource has often been solely associated with developing countries. However, Aguilar (2015) stresses the importance of wood in developed nation's energy markets, specifically in the growing trend of mandated transitions to more renewable energy resources. In the United States alone, wood energy provides 25% of renewable energy consumption, which is greater than both wind and solar energy. As wood energy is often overlooked, he highlights the importance of recognizing this valuable and complex resource.

The EU has almost doubled its wood energy consumption since 1990 largely due to the "20-20-20 by 2020" mandate, which states that 20% of the energy in participating countries must come from a renewable energy source, along with a 20% increase in energy efficient and a 20% decrease in greenhouse gas emission, all by the year 2020. Because of this, the EU is also expected to double its demand for wood energy within the next 20 years and will depend heavily on imports from other countries, such as the United States and Canada, to acquire such resources. Additionally, wood energy is accepted under many renewable portfolio standards that have been implemented in the United States. These factors display the importance of understand the wood energy market, as it is likely to boom in upcoming years.

One reason why wood energy is likely to boom is due to the relatively easy transition to wood energy. Wood has become a worldwide market, as seen with the growing demand for imports to the EU as mentioned above. This has been facilitated by the "pelletization" of wood, which allows the energy found in wood fuels to be condensed, creating a more cost-effective way to transport wood. Additionally, altering power plants for wood energy is a relatively small investment, especially as wood is very price-competitive to other fuel sources.

As more developed nations switch to renewable energy resources, it is very likely that wood energy will be a prominent choice due to the relatively easy transition. It is important to recognize this, as many interests are involved regarding wood energy such as land management, energy production, and environmental objectives, all of which need to be considered. Without recognizing the pivotal point wood energy plays in the developed nations' economies and with many players involved, it is likely that management will not be carried out effectively, and therefore limit the capability that wood energy has to provide to the energy market.

Aguilar, F., 2015. Wood Energy in Developed Economies: An Overlooked Renewable. Resources for the Future 188, 22-27.

Does Biofuel Have a Viable Future?
by Ali Siddiqui

An article by Justin Gillis for The New York Times discusses a new report published by World Resources Institute, a global research organization based in Washington, that suggests that biofuels are not the direction policy makers should be heading when considering alternative energy. Biofuels are fuels created by plant matter. Timothy D. Searchinger, a research scholar at Princeton and primary author of this report, was quoted in this article to have said that they were an "inefficient way to convert sunlight to fuel".

The report has stated that earlier policy made under the assumption that biofuels were a shining promise in the future of energy has actually helped drive up global food prices, worsened some types of air pollution, and done little to reduce carbon emissions. The thought was that these biofuels would be grown and burned, but that the new crops put in place would remove the released emitted gas before being burned again. This idea in practice fell short.

Another interesting consideration put forth was that the efficiency of biofuels with respect to how it incorporated the value of its land was missing. The opportunity cost of what else could be done with the land is significantly higher in the minds of some scientists than the recent low success of biofuels.

Also interesting to note was that most of the pro-biofuel policies that were adapted were done in a period of time when solar energy like many other sources of renewable energy was extremely expensive. In a time, when the costs of these other potential sources of energy have slightly dropped, is it time to drop biofuels as well?

Gillis, Justin. New Report Urges Western Governments to Reconsider Reliance on Biofuels. New York Times. January 19th 2015.
http://www.nytimes.com/2015/01/29/science/new-report-urges-western-governments-to-reconsider-reliance-on-biofuels.html?ref=science&_r=1

Pee-Power Could Save Lives
by Abigail Wang

Human waste might be the next hot commodity. With the biomethane gas-powered "Bio-Bus" recently rolled out in the United Kingdom, human waste is rapidly becoming an area of interest for energy enthusiasts. We have one use for fecal waste, but what about fluid waste? Researchers at University of West England (UWE Bristol) working together with Oxfam came up with a solution: a urinal that can generate electricity.

Cleverly placed by the Student Union Bar at UWE Bristol, the urinal is routed to microbial fuel cell (MFC) stacks, fixed at the back of the cubicle, that generate enough electricity to power indoor lighting. When urine flows through an MFC, the microbes use it as part of their metabolic process, which frees electrons in the process. The electrodes inside the cell gather the electrons and when connected to a circuit, generate a current. This process is dubbed urine-tricity or pee power.

The project is being led by Ioannia Ieropoulus, Director of Bristol BioEnergy Centre, who claims that this could be a game-changer for refugee camps by bringing light to their bathroom facilities. Refugee camps are often dark and dangerous places, especially for women who are at risk for assault and harassment. The first toilet is going to be sent to a refugee camp within the next six months. Once testing is completed, the technology will be rolled out first to refugee camps and then to other places without electricity. The Bill & Melinda Gates Foundation is funding the urine-tricity project while the Engineering and Physical Sciences Research Council is funding the Microbial Fuel Cell work.

The BioEnergy Team from Bristol Robotics Laboratory, which is a collaboration between UWE Bristol and the University of Bristol in the UK, has already proven that this method of generating electricity works. In 2013, the group showcased a smart toilet at the Re-Invent the Toilet Fair: India. The fair, put on by the Indian Ministry of Urban Development and the Bill & Melinda Gates Foundation, aims to stimulate discussions and encourage partnerships to improve global sanitation and bring affordable sanitation solutions to areas in need. In the demonstration, the researchers showed how the electricity generated by microbial cell stacks could power mobile phones. The technology is also capable of removing pathogens and cleaning the urine for fertilizer and sanitation purposes.

The technology seems sustainable in the long-term because the only thing it needs is a plentiful supply of urine. Researchers estimated that humans produce about 6.4 trillion liters of urine across the globe every year, proving that liquid waste is an abundant resource with lots of potential. The technology itself is also not too expensive; one microbial fuel cell costs about two dollars and to set up a small unit in a camp would cost a little over $900. These toilets also don't have to only be used in developing countries; Ieropoulos hopes that these microbial fuel cells can be retrofitted to any existing toilet and take the flushed waste to generate electricity as it goes down the drain.

"BRL to be Exhibitor at Reinvent the Toilet Fair: India to Showcase Advancements that Improve Sanitation and Health." University of the West of England: 18 March 2014. http://info.uwe.ac.uk/news/uwenews/news.aspx?id=2777

"Urine Power to Light Camps in Disaster Zones." ScienceDaily: 6 March 2015. http://www.sciencedaily.com/releases/2015/03/150306111905.htm?utm_source=feedburner&utm_medium=feed&utm_campaign=Feed%3A+sciencedaily%2Fmatter_energy%2Fenergy_technology+%28Energy+Technology+News+--+ScienceDaily%29

Mis, Magdalena. "Pee-powered 'Green' Toilet to Light Up Refugee Camps." Business Insider: 4 March 2015. http://www.businessinsider.com/r-pee-powered-green-toilet-to-light-up-refugee-camps-2015-3

Gas-to-Liquids Technology
by Alex Elder

Siluria Technologies, a San Francisco-based company, has recently developed a technology that readily converts natural gas to gasoline and diesel byproducts. The process works by using a chemical catalyst to convert ethylene gas into liquid form. Siluria named this process "ETL" or Ethlylene to Liquids. Although other companies have developed methods of converting gases into liquid energy, Siluria's method differs substantially from other processes in that their process does not go through any carbon monoxide and hydrogen intermediates (or "syngas"). Having syngas in the conversion process requires significant additional refining and energy input before the final product is reached. In contrast, using ethylene has as an intermediate in place of syngas allows for a targeted production of the specific desired product of gasoline and diesel. The overall result of this innovative method is a simpler process, with lower costs and more flexibility in terms of scale.

This less expensive way to turn natural gas into motor fuels could also potentially benefit consumers because natural gas is far cheaper and more plentiful than crude oil. With their revolutionary conversion process, gasoline could be far more affordable to the average consumer. Siluria has been testing this technology at facilities in California for several years and recently brought online a demonstration plant in La Porte, Texas. The company aims to commercialize it by 2017.

Hays, K. (2015). "California energy tech firm pushing gas-to-liquids technology." Reuters. http://www.reuters.com/article/2015/04/01/siluria-natgas-gasoline-idUSL2N0WW2KI20150401
Siluria Technologies. Siluria.com

GreatPoint Energy Makes a Great Point

by Alexander Flores

GreatPoint Energy has discovered a way to produce low-cost, clean natural gas from biomass, petroleum coke, and coal utilizing its patented Bluegas catalytic hydromethanation process. This Bluegas is able to operate at a much higher efficiency than competing technologies while benefitting from lower capital intensity, a superior environmental footprint, nearly complete carbon capture, and a significantly lower cost of production. The natural gas produced can be transported through existing pipeline infrastructure while remaining much less expensive than liquefied natural gas. By being completely interchangeable with drilled natural gas, it also can be used for power generation, residential and commercial heating, and the production of chemicals.

So how exactly does the Bluegas catalytic hydromethanation process work? Essentially, this process entails carbon-rich feedstocks being converted through a combination of catalytic reactions in the presence of water into a methane-rich gas stream. First, a catalyst is dispersed throughout the matrix of a carbon-rich feedstock under specific conditions to ensure effective reactivity. The catalyst/feedstock material is then loaded into a hydromethanation reactor where pressurized steam is injected to fluidize the mixture and ensure continuous contact between catalyst and carbon particles. The catalyst then facilitates multiple chemical reactions between the carbon and the steam on the surface of the particles. The overall combination of reactions that occur are thermally neutral, which make it highly efficient. The sulfur-tolerant catalyst formulation is composed of abundantly available, low cost metal materials designed to promote gasification at the low temperatures where methanation and water-gas shift occur. This catalyst is then recycled and reused within the process. The addition of such a catalyst allows the reduction of operating temperatures in the gasifier while promoting yields of methane. With mild catalytic conditions, less expensive reactor components are used along with low-cost carbon sources, while pipeline grade methane is produced. Overall, the Bluegas technology permits the recovery of contaminants in coal, petroleum coke, and biomass as useful byproducts. Additionally, nearly all of the carbon dioxide produced is captured as a pure stream, which is suitable for sequestration or enhanced oil recovery. The company intends to develop, own, and operate large-scale Bluegas production facilities with local partners in the near future. The full-scale facilities will be strategically located in areas with the greatest price differential between delivered feedstock and natural gas markets for maximum effectiveness. Great point, GreatPoint.

GreatPoint Energy 1 (http://www.greatpointenergy.com/about.php)
GreatPoint Energy 2 (http://www.greatpointenergy.com/ourtechnology.php)
GreatPoint Energy 3 (http://www.greatpointenergy.com/clearadvantages.php)
GreatPoint Energy 4 (http://www.greatpointenergy.com/environmentalprofile.php

Waste and Fuel with InEnTec
by Alexander Flores

InEntec has developed a proven solution to aid our world with the confounding challenges of discovering new sources of clean fuels for electricity and transportation and disposing of the millions of tons of waste in an environmentally-responsible manner. Using their patented Plasma Enhanced Melter (PEM) technology, virtually any waste—hazardous, medical, industrial, and municipal waste—can be converted into clean energy products that can be used for transportation fuels, industrial products, and electricity generation. The Plasma Enhanced Melter (PEM) utilizes plasma power to break down materials into their elemental components (hydrogen, carbon, oxygen, etc.) and then transforms them into synthesis gas, which is a basic building block of the chemical and energy industries. How is this done exactly? First, feedstock is introduced into the PEM process via a feed system specifically designed for the type of materials being processed. This is followed by a "preliminary" processing zone where the Pregasifier converts approximately 80% of the organic feedstock portion into syngas. Remaining feedstock, consisting of inorganic materials, carbon, and additional unprocessed organics pass through outlet of the Pregasifier and into the PEM Process Chamber. In the PEM Process Chamber, the remaining feed materials from the Pregasifier are dropped onto a molten glass surface near a plasma-arc zone, which is DC powered. This plasma arc provides enough intense energy required to quickly gasify remaining organic materials for their conversion into syngas, while remaining inorganic components are incorporated into the molten glass bath, which is AC powered. The created syngas exits the PEM Process Chamber and then flows into the Thermal Residence Chamber (TRC). The Thermal Residence Chamber provides additional residence time at a high enough temperature to completely process remaining organic materials in the syngas in order for the gasification to reach equilibrium. Drain power supplies are then used to operate the drains, allowing molten glass and metal to be removed from the system. This is followed by syngas cleaning through a series of standard process to prepare it for final product use. Ultimately, InEnTec's PEM system is an amazing solution to provide lower waste processing costs, optional on-site power generation, greater recycling of waste materials into commercial products, reduced Greenhouse Gas emissions, and eliminated future liability for the waste. So, who's in with InEnTec?

InEnTec 1 (http://www.inentec.com/)
InEnTec 2 (http://www.inentec.com/about-inentec.html)
InEnTec 3 (http://www.inentec.com/pemtm-technology.html)
InEnTec 4 (http://www.inentec.com/pemtm-technology/process-details.html)
InEnTec 5 (http://www.inentec.com/videos.html)

HYDROGEN

Gaseous hydrogen is a useful fuel because it can be combusted in a fuel cell to make electricity with no byproducts other than heat and water. But it has to be manufactured, usually either by stripping it off of natural gas, in which case it has the same greenhouse gas and air pollution effects of burning the natural gas directly, albeit in a potentially remote location, or by utilizing electricity to separate it from water. If the electricity comes from renewable energy, then hydrogen becomes an extremely clean fuel. The main barriers to its replacing gasoline as a vehicle fuel are its relatively low power density—we haven't figured out how to effectively compress it very much—,the lack of pipeline infrastructure for delivering it to end users, and a lack of small, relatively low temperature, affordable fuel cell technology suitable for vehicles. All of these barriers seem fully amenable to engineering solutions, so it seems likely that hydrogen-fueled vehicles will gradually penetrate the world's transportation fleets, starting with larger commercial vehicles since technology for larger fuel cells is more mature.

Toshiba Opens Hydrogen Energy Research Center in Tokyo
by Mariah Valerie Barber

On April 7th 2015, Toshiba opened a large Hydrogen Energy Research Center in the Fuchu Complex in Tokyo. This research center will aim to develop the necessary technology to make hydrogen-related energy technology part of the energy economy in Tokyo. Construction for the 900 square-meter, light-gauge steel center began in December 2014 and ended with the opening. Toshiba will be working to install solid oxide electrolysis cell (SOEC), solar photovoltaic generation systems, and fuel cells within it (Clean Energy). Toshiba aims to produce hydrogen using such systems. Hydrogen is considered to be a very clean fuel source. In addition, the hydrogen will be created utilizing generation systems that come from renewable energy sources such as wind, hydro, and photovoltaics. Once the hydrogen is produced Toshiba aims to then use fuel cells to convert it into electricity, establishing a stable electricity supply in Japan. Toshiba hopes to be able to develop hydrogen-generating water electrolysis systems that will be able to be installed in isolated areas or islands to provide steady access to electricity. Additionally, the Toshiba press release stated, "Output from solar and wind power sources is unreliable, but it can be used to power water electrolysis systems that produce hydrogen, and that hydrogen can be transported and stored for use when needed" (Toshiba).

Toshiba is aiming to increase the hydrogen related electricity and energy business in Japan up to $800 million by 2020 and $4.10 billion dollars by 2030. The company is aiming to make itself one of the main suppliers in hydrogen by 2025, selling its hydrogen internationally. It will also use in the hydrogen-fired gas turbines of Japan (Clean Technology).

Clean Technology (http://www.cleantechnology-business-review.com/news/toshiba-opens-hydrogen-energy-research-center-in-japan-070415-4547837)
Toshiba (http://www.toshiba.co.jp/about/press/2015_04/pr0601.htm)
Bloomberg Business (http://www.bloomberg.com/news/articles/2014-12-09/japan-promotes-home-fuel-cell-on-path-to-hydrogen-society)

The Better Way to Produce Hydrogen Fuel
by Nour Bundogji

A team of Virginia Tech researchers developed a method to create hydrogen fuel from biomass in a timely and financially efficient manner. Other biomass fuel production methods rely on highly processed sugars to create their fuel but their method uses abundantly available corn stalks, cobs, and husks (aka corn stover) to produce hydrogen. A commentator at ScienceDaily.com explained that their method "not only reduces the initial expense of creating the fuel, it enables the use of a fuel source readily available near the processing plants, making the creation of the fuel a local enterprise."

Furthermore, the team's new findings could help speed the widespread arrival of hydrogen-powered vehicles in a way that is inexpensive and has extremely low carbon emissions.

Their new discovery is unique in two ways: first, the use of dirty biomass (the husks and stalks of corn plants) to create their fuel not only reduces the initial expense of creating the fuel, but also allows for the use of a fuel source readily available near the processing plants. Joe Rollin, a former doctoral student at

Virginia Tech, used a genetic algorithm and a series of complex mathematical expressions to analyze each step of enzymatic processes that break down corn stover into hydrogen and carbon dioxide. The largest hurdles to widespread hydrogen use are the capital cost required to produce the fuel from natural gas in large facilities and distribution of hydrogen to fuel cell vehicles. However, Rollin's model increased reaction rates by threefold, decreasing the required facility size and reducing associated capital costs. In terms of product yield, their method "not only breaks the natural limit of hydrogen-producing microorganisms by three times but also avoids complicated sugar flux regulation." Furthermore, the team also increases enzymatic generation rates to at least 10 times that of the fastest photo-hydrogen production system.

The team already has received significant funding for the next step of the project, which is to scale up production to a demonstration size.

"Although it is difficult to predict cost at this point, this work represents a revolutionary approach that offers many new advantages," said Lonnie Ingram, director of the Florida Center for Renewable Chemicals and Fuels at the University of Florida, who is familiar with the work but not associated with the team. "These researchers have certainly broadened the scope of our thinking about metabolism and how it plays into the future of alternative energy production."

Joseph A. Rollin, Julia Martin del Campo, Suwan Myung, Fangfang Sun, Chun You, Allison Bakovic, Roberto Castro, Sanjeev K. Chandrayan, Chang-Hao Wu, Michael W. W. Adams, Ryan S. Senger, and Y.-H. Percival Zhang. High-yield hydrogen production from biomass by in vitro metabolic engineering: Mixed sugars coutilization and kinetic modeling. PNAS, 2015 DOI: 10.1073/pnas.1417719112

Virginia Tech. "Discovery may be breakthrough for hydrogen cars." ScienceDaily. ScienceDaily, 7 April 2015. <www.sciencedaily.com/releases/2015/04/150406152955.htm>.

Hydrogen Hungry Bacteria Bring Artificial Leaf One Step Closer to Viability

by Liza Farr

In 2011, Daniel Nocera engineered an artificial leaf that uses only sun and water to produce energy (Chandler, Sep 30, 2011). The leaf was made of silicon solar plates with different catalytic materials bonded on each side (Chandler, Sep 30, 2011). When the plate is placed in water and exposed to sunlight, one side produces hydrogen bubbles, and one side oxygen bubbles, which can be stored and used for energy

(Chandler, Sep 30, 2011). Although this was an important innovation in renewable energy, major shortcomings of the invention was that it produced hydrogen, which does not easily fit into our existing energy infrastructure, rather than liquid fuel. Recently, Nocera has collaborated with biologists at Harvard University to engineer bacteria that convert hydrogen into an alcohol-based fuel (Nunez, Feb 9, 2015).

In natural photosynthesis, biomass is produced when sunlight meets water and carbon dioxide, but an extra step is needed to turn the biomass into fuel, such as the fermenting of corn to make ethanol, which can be energy intensive and inefficient (Nunez, Feb 9, 2015). The new bacteria bypass this step to produce liquid fuel immediately, by absorbing the hydrogen and combining it with carbon dioxide to produce isopropanol, an alcohol-based fuel similar to ethanol (Nunez, Feb 9, 2015). In order to make this bacteria economically viable, researchers needed to produce a 10% solar-to-fuel conversion efficiency rate, which they achieved and published in a scientific article in August, 2014 (Coxa *et al.* 2014). Both the leaf and the bacteria catalyst are made using non-precious, earth-abundant materials, bringing the project another step closer to economic viability (Coxa *et al.* 2014). Nunez (Feb 9, 2015) predicts that the next step needed to make this innovation commercially viable is to figure out how the carbon dioxide required to complete the reaction can be obtained directly from the atmosphere. This process is quite energy intensive, and it is a major barrier to the viability of the innovation. This will undoubtedly be the next problem Nocera and others will address to make the artificial leaf a viable innovation (Nunez, Feb 9, 2015).

Casandra R. Coxa, Jungwoo Z. Leeb, Daniel G. Noceraa, and Tonio Buonassisi. 2014. Ten-percent solar-to-fuel conversion with nonprecious materials. Proceedings of the National Academy fo Sciences. August 1, 2014:14057–14061.

Chandler, David L. 'Artificial leaf' makes fuel from sunlight. September 30, 2011. [http://newsoffice.mit.edu/2011/artificial-leaf-0930]

Nunez, Christina. "Tweaking Bacteria, Scientist Turn Sunlight into Liquid Fuel." February 9, 2015. [http://news.nationalgeographic.com/news/energy/2015/02/150209-solar-energy-to-liquid-fuel/]

Ideal Shift from Gasoline to Hydrogen Fuel Cells

by Alexander Flores

The main alterative for gasoline fuel and battery electric vehicles is one involving the utilization of hydrogen fuel cells. Just how do these hydrogen fuel cells work? Essentially, each fuel cell is an anode and cathode with a proton exchange membrane sandwiched in between. Hydrogen from an onboard tank would enter the anode side of the fuel cell, while oxygen in the atmosphere would enter the cathode side. Once the hydrogen molecule encounters the membrane, a catalyst forces its split into proton and electron. The proton would then move through the fuel cell stack as the electron follows an external circuit, delivering an electric current to the motor and other parts of the vehicle. The proton and electron would join again at the cathode side and combine with oxygen to form water as the main emission.

This fascinating science and technological application has many automakers relieved since sales of electric cars and plug-in hybrids are slow. Even Toyota, creator of the Prius gas-hybrid, will opt to utilize hydrogen fuel cells rather than batteries to power its next generation of green vehicles. The main issue of concern regarding hydrogen-powered cars is that they are expensive, just like electric cars. This is why we've yet to see such vehicles on the market, especially due to the lack of fueling stations for them.

One of the few early investors in hydrogen stations, Dan Poppe, claims that a much higher number in sales of hydrogen fuel cell vehicles needs to be reached prior to the building of more fueling stations. Fueling stations would allow a driver to refill his or her hydrogen tank within minutes versus having to recharge a battery pack for hours. Unfortunately, according to Poppe, there are approximately 250 hydrogen fuel cell vehicles on the road to date and that just won't cut it for exponential growth. So, California intends to have 1.5 million zero-emission cars on the road by 2025 along with 15% of all new cars being sold to be zero-emission vehicles.

At this point, automakers will be rewarded more environmental credits by California for building hydrogen fuel cell cars than for battery electric cars or plug-in hybrids. Dan Poppe is currently feeling the pressure after receiving a $3 million from the state to build a station in Chino along with $500,000 from the energy commission and air quality district to operate a station in Burbank. In order to receive grants to cover operational expenses, Poppe must reach specific performance goals – specific number of pumps open, operation at certain capacities, by specific dates—or face disqualification of his projects.

Toyota intends to launch a hydrogen fuel cell sedan in Japan early this year and in the US by the summer, while Hyundai has already began leasing a hydrogen fuel cell version of its Tucson sport vehicle. Honda, Ford, and GMC all intend to be a part of this new technology as well to ensure progression and ultimately the greenest vehicle possible. Since fuel cell cars have approximately the same range as many gas-powered vehicles (300 miles), it is quite easy to see why some of the biggest automakers may just turn away from the typical 80-mile range, battery-powered cars.

California's state Legislature passed AB 8 last year and has dedicated $20 million a year through 2023 to finance the construction of approximately 100 hydrogen fueling stations. If automakers and scientists are able to maximize sources of hydrogen (hydro-electric or wind generators, nuclear power plants, natural gas) then this would be revolutionary to our world of transportation.

Drive Clean
(http://www.driveclean.ca.gov/Search_and_Explore/Technologies_and_Fuel_Types/Hydrogen_Fuel_Cell.php)
Fleming, Charles. "Carmakers prepare to shift to hydrogen fuel cells." Los Angeles Times. October 26, 2014.
Los Angeles Times (http://www.latimes.com/business/autos/la-fi-hy-fuel-cell-cars-20141026-story.html#page=1)

H2FC SUPERGEN: UK Hydrogen and Fuel Cell Research at its Finest
by Chad Redman

High on the list of fuels with the potential to reduce humankind's carbon footprint sits an abundant resource, hydrogen. Given such potential, the Hydrogen and Fuel Cell (H2FC) SUPERGEN Hub in the United Kingdom focuses on research and development for the hydrogen and fuel cell industry. The Hub is a multifaceted organization launched in May 2012 that combines scientific, economic, and policy-related R&D under one roof. Funding from the UK's Engineering and Physical Sciences Research Council draws in professionals from all three camps and encourages collaboration with the goal of advancing hydrogen and fuel cell use across the globe.

The H2FC SUPERGEN Hub is compartmentalized into several different divisions, each with its own specialization in the hydrogen and fuel cell industry. For example, one division of the Hub is the Systems group, focusing on integrating separate hydrogen and fuel cell units, as well as large system optimization. The research conducted by the Systems group includes topics such as developing a refueling supply chain.

At the other end of the Hub's research spectrum is the Socioeconomic research group. This group has published research on the socioeconomic feasibility of technologies such as hydrogen fuel cell cars, covering the economic and political aspects of the hydrogen and fuel cell industry. Considerations such as infrastructure cost and policy challenges are handled by the socioeconomic research group.

As an organization, the UK's Hydrogen and Fuel Cell SUPERGEN Hub is a sign that hydrogen may become a common source of energy in the not-so-distant future. Collaboration and funding for projects such as these give high hopes to those seeking more responsible ways to power our lives.

Chloe Stockford, Nigel Brandon, and John Irvine. "H2FG SUPERGEN: An Overview of the Hydrogen and Fuel Cell Research Across the UK." International Journal of Hydrogen Energy 40 (2015): 5534-543. Web. 30 Mar. 2015.

Feeding Cars With Plants
by Abigail Wang

Who knew you could feed your car cornhusks? Thanks to researchers at Virginia Tech, cornhusks might be fueling our cars in the near future. They've developed a cheaper and less time-consuming way to create hydrogen fuel using the husks, cobs, and stalks of corn plants.

Hydrogen fuel is a green fuel that has nearly zero carbon emissions. Current methods to produce hydrogen fuel are expensive, use highly processed sugars, and take a long time. However, Percival Zhang, a professor in Virginia's Tech Department of Biological Systems Engineering, and Joe Rollin, a former doctoral student at Virginia Tech, found an easier way to produce hydrogen fuel. They published their findings in the journal Proceedings of the National Academy of Sciences (PNAS).

The overall process involves mixing the cornhusks with a special solution made up of ten enzymes that turn plant sugars into hydrogen and carbon dioxide. The enzymes were created in microbial fermenters with genetically engineered bacteria. After the enzymes are added to the plant waste, the solution is left for several weeks.

Rollin used a genetic algorithm with series of complex mathematical expressions to look at every step of the enzymatic process that breaks down the discarded corn parts into hydrogen and carbon dioxide. He was able to make this new method utilize glucose and xylose simultaneously, which increases the rate at which hydrogen is released. With this new method, the system for creating hydrogen fuel tripled reaction rates and only requires a facility the size of a gas station. This is much smaller than current hydrogen fuel production facilities.

It used to only be possible to convert 30% to 60% of plant sugar into hydrogen using industrial catalysts or fermentation, now it's feasible to convert 100% of sugar in corn stalks and husks into hydrogen gas. Since it's also cheaper to produce fuel this way, processing plants that make corn-based products could start to make fuel to power their own operators.

Zhang and Rollin created start-up Cell-free Bioinnovations to work on their latest finding. They hope in the next three to five years they will build a bioreactor that could produce 200 kilos of hydrogen fuel a day, which would refuel 40 to 50 cars. This innovation is an important step toward a hydrogen economy.

Griffiths, Sarah. "Hydrogen Fuel Breakthrough as Plant Waste is Converted Into Gas with 100% Efficiency." DailyMail.com: 6 April 2015. http://www.dailymail.co.uk/sciencetech/article-3027932/Corn-husks-promising-source-renewable-fuel.html

Russon, Mary-Ann. "Hydrogen Fuel Made From Corn Husks Could be Renewable Energy Breakthrough for Cars." International Business Times: 7 April 2015. http://www.ibtimes.co.uk/hydrogen-fuel-made-corn-husks-could-be-renewable-energy-breakthrough-cars-1495253

HYDRAULIC FRACTURING AND CARBON SEQUESTRATION

Hydraulic fracturing (fracking) is freeing up previously inaccessible natural gas and oil supplies, leading to at least two beneficial outcomes: decreasing dependence on fossil fuel imports for the US, and lower carbon dioxide emissions as natural gas replaces coal as a fuel for electrical generating stations. There are some downsides as well; industrialization of previously bucolic farmland, contamination of water supplies, increased seismic activity, increased air pollution from wellheads. Is it worth it? This section examines some of the research examining the pros and cons.

Why Fracking Works (and Sometimes Doesn't)
by Emil Morhardt

We hear a great deal about the economic benefits of hydraulic fracturing, and even more about its potential liabilities, but seldom very much about exactly how fracking works. A fascinating paper published by the American Society of Mechanical Engineers (Bazant *et al.* 2014) combines an extremely clear explanation of the process in non-technical language with a detailed mathematical analysis of the mechanics involved (a combination uncommon in engineering papers). The question at hand is why, with pipes just three-inches in diameter, spaced half a kilometer apart, it is possible to get so much gas out of shale beds. The first thing to know is that even this technology gets only about 5–15% of the gas embedded in the shale, so it's likely they'll be going back for more as the technology improves. They know about this percentage because of how much gas they can extract from the rock samples they get out of the well cores.

The reason fracking is useful in the first place is that the gas, in the form of solid kerogen, is trapped in nanovoids, and the natural cracks that would let it migrate to a borehole are either squeezed shut by the weight of the three kilometers of rock above them, or filled up with calcite or other minerals. The trick is to open up the existing cracks, form new ones, then keep them open so gas can flow out. Through a lot of detailed engineering calculations the authors determine that the way to open up the most cracks (and in the process to keep the most fracking fluid stuck in the shale, rather than returning to the surface where it needs to be treated or re-injected) is to gradually increase the hydraulic pressure to a point where it gradually opens up vertical cracks, while the sand they also inject gradually fills an ever-widening array of them. The acid included in the fracking fluid apparently breaks up the rough edges (asperities) of the cracks into pieces small enough to act like the sand, increasing the amount of crack propping. Since the injection pressure of the fracking fluid seldom exceeds the pressure exerted by the overburden, horizontal cracks in the natural bedding plane are equally seldom opened up. The whole procedure at this point in its evolution is dependent on vertical cracks.

Methane, which makes up most of the content of natural gas, does not dissolve in water (nor in fracking fluid, which is 99% water) so it migrates through the cracks and into the well casing in the form of gas bubbles, which, when they reach the vertical borehole are aided in coming to the surface because of their buoyancy.

The graphs included in the paper show that the highest rate of gas flow occurs soon after a well is drilled, with exponentially decreasing flow over the four- or five-year life of the well.

So, in summary, fracking usually occurs in shale layers 20 to 150 meters thick, lying on the order of three kilometers below ground—a long way below any aquifers that might be susceptible to contamination, but of course the well shaft must pass through these aquifers. From each vertical borehole, several horizontal ones several kilometers long are drilled about 500 meters apart. These horizontal holes are lined with three-inch diameter steel pipe that is then perforated explosively at five to eight locations along their lengths. Several million gallons of water with a little sand, acid, and other chemicals is injected under great pressure. (Although this sounds like a lot of water it corresponds to only 1-2 millimeters of rain falling over the well field.) The injection initially opens up natural vertical cracks in the shale, which are typically 15 to 50 centimeters apart. Continued pressure increases the array of vertical cracks, ideally about 10 centimeters apart. The methane forms bubbles and flows along the pressure gradient toward the horizontal pipe, then through the pipe to the vertical borehole, and up to the surface to be captured.

If you're not familiar with materials science and engineering technical writing, I'd suggest looking at the full paper. It opens a window into the thought processes of engineers that most of us never encounter.

Bazant, Z.P., Salviato, M., Chau, V.T., Viswanathan, H., Zubelewicz, A., 2014. Why Fracking Works. Journal of Applied Mechanics 81.

Why Fracking Might Not Work for as Long as We Would Like
by Emil Morhardt

The December 4, 2014 issue of the scientific journal Nature takes the position that the current abundance of natural gas in the US derived from horizontal drilling and hydraulic fracturing may be a much shorter-term phenomenon than most analysts have thought. In both an editorial and an opinion piece (not however in a scientific paper) the journal takes issue with the US Energy Administration's (USEA) assessment that natural gas production in the US will continue to grow for a quarter century, at least. Nature relies on the opinions of a team of researchers at the University of Texas, and cites a paper (Patzek, 2012) by members of the team, which now consists of a dozen geoscientists, petroleum engineers, and economists. That paper examines extraction data from 2,057 such wells in the oldest US shale play, the Barnett Shale in Texas, and concludes that they started to decline at an exponential rate in ten years or less, and goes on to predict the total amount of gas that will be produced by their overall sample of 8,294 wells; 10–20 trillion standard cubic feet over the next 50 years.

The team producing the paper talked to Nature's Mason Inman (2014) about current work that is only beginning to appear at conference presentations and in scientific journals, but which suggests that the major current US shale gas operations would peak in 2020, and decline from then on, producing only half as much gas by 2030 as predicted by the USEA, even under its most conservative scenarios. This discrepancy may be attributable to the more detailed look at producing wells taken by the Texas team, and the USEA is likely to do similar analyses itself. The bottom line, though, seems to be an increase in uncertainty (not a big surprise to anyone who has done scientific research—the more closely one looks at a problem, the more complicated it becomes, usually.) So one might conclude that the research should serve to temper the euphoria on the part of those profiting from the lower gas prices, as well as the frustration felt by producers as they watch gas prices fall, and by environmentalists who fear that the low prices will stifle attempts to replace fossil fuels with renewables.

Patzek, T.W., Male, F., Marder, M., 2013. Gas production in the Barnett Shale obeys a simple scaling theory. Proceedings of the National Academy of Sciences 110, 19731-19736.

Tracking Fracking Fluid with Molecular Tracers
by Emil Morhardt

Stephanie Kurose, a law student at the American University in Washington DC, calls our attention to both the concept of, and two startups trying to push, micro-tracers which could be injected into fracking fluid so that if it escapes, we know whodunit. The idea is simple, if not yet operational; create some long-lived non-toxic chemical compound with enough potential variation that a different version could be mixed in with the fracking fluid for each individual well. The arguments for it, espoused by Kurose, are equally simple; drilling companies would know if they had a problem with leakage and could change their technology, falsely-accused drilling companies could exonerate themselves, and the public should feel much less angst about fracking if evidence of leaked fracking fluid fails to materialize (or vice versa.) It might be that drilling companies would resist in order to avoid any conclusive evidence that their wells have leaked, but so far no one knows because suitable tracers have yet to be deployed. The two startups giving it a shot are BaseTrace and FracEnsure.

BaseTrace uses genetic engineering technology to produce strands of resilient DNA, which can be readily customized into a nearly infinite number of variations which could be mixed with all sorts of industrial fluids, including fracking fluid. Genetic engineering technology makes it equally simple to read the genetic code in these relatively short, by biological standards, strands of DNA.

FracEnsure uses nanoparticles with a paramagnetic coding that is somehow individually coded in batches, but the company's website does not explain the technology further so we will have to wait.

Kurose, S., 2014. Requiring the Use of Tracers in Hydraulic Fracturing Fluid to Trace Alleged Contamination. Sustainable Development Law & Policy, Summer 2014, page 43.

Instead of Flaring Natural Gas at Fracked Oil Wells, Use it to Treat Fracking Fluid

by Emil Morhardt

Seems like a good idea. Yael Rebecca Glazer suggested it in a 2014 Masters Thesis in Engineering at the University of Texas at Austin. A major issue with fracking is that sometimes a lot of the fracking fluid that was pumped down the well to create the fractures comes back up, sometimes along with additional "produced" water, sometimes twice as much as was pumped down in the first place. On top of that, it is often so contaminated that it exceeds the capabilities of industrial treatment facilities, so it gets trucked to a nearby injection well and is reinserted. But injection wells are not always handy, and anyway, the water itself would be valuable if it weren't so polluted. Meanwhile, although a fracked well might producing mainly oil, there is also often a fair amount of natural gas produced; but if there isn't enough gas to make it economical to capture it and sell it, it is commonly flared—burned right there at the wellhead. This converts the natural gas to CO_2 without using the energy released for anything at all. Maybe, thought Ms. Glazer, that free energy could be used onsite to power wastewater cleanup technologies that normally wouldn't be considered because of their high energy costs. It also occurred to her that since lots of these wells are in the sunny, windy southwestern US, local photovoltaic panels or wind turbines might supply energy as well. This latter option is attractive when there are no convenient transmission lines to take the power offsite, even though solar or wind energy is abundant.

So the first idea is to channel the heat from burning the otherwise flared gas to heat-powered water treatment technologies like multi-stage flash distillation, multi effect distillation, and mechanical vapor recompression. Heat could also come from a solar thermal facility. Alternatively, if not enough gas were available for producing the heat, electricity from photovoltaics or wind turbines could pressurize the wastewater for membrane separation, or the related reverse osmosis. Reverse osmosis is the most energy-efficient of the treatments considered here, so it would be preferable if energy were limiting.

Ms. Glazer uses a series of equations to do a formal engineering analysis of the feasibility of her suggestions, and figures they could reduce overall water requirements for fracking by 11–26%, and reduce the energy use for freshwater trucking by 16%. If renewable energy were used there would need to be at least four 100 kW wind turbines (much smaller than the ones we are accustomed to seeing at wind farms) or between 1000 and 4000 250 Watt PV panels. She thinks she's on to something, and I'd agree. But even though the fuel is free, the equipment and maintenance surely are not, so this probably won't happen without a mandate, and unless there's a full-scale demonstration of the feasibility and cost, no mandate is likely either. Maybe DOE or EPA should fund a demonstration project.

Glazer, Y.R., 2014. The potential for using energy from flared gas or renewable resources for on-site hydraulic fracturing wastewater treatment. Thesis, Master of Science in Engineering, Graduate School. University of Texas at Austin, 83 pages.

Using Supercritical CO_2 Instead of Water for Fracking
by Emil Morhardt

The purpose of hydraulic fracturing is to use high pressure to open up pores in deep fuel-bearing shale deposits so that the oil or natural gas can escape through boreholes to the surface. To make this work, very high pressures (hence, much surface equipment) and a great deal of water are required. To keep the pores propped open when the pressure and water recede, something (usually sand) needs to be included. The inclusion of acid can increase pore efficiency, and because water is a good biological medium, antibacterial agents may be required to prevent fouling. Finally, most of the fracking fluid returns to the surface where it presents a treatment and disposal problem. But in theory, any liquid, or supercritical substance, would work, supercritical CO_2 (sCO_2), for example. According to a study underway at Los Alamos National Laboratory (Middleton *et al.* 2014) sCO_2 has a number of potential advantages over water, and some potential disadvantages as well.

The advantages are striking; it requires less pressure (so less equipment) at the well pad, it displaces gas from lower-porosity fractures and mobilizes it from organic inclusions, it mobilizes heaver hydrocarbons, it doesn't need the additives—maybe not even the proppant—now included in fracking fluid, and, perhaps best of all, it competitively displaces methane from the shale, preferentially absorbing on to it, staying in the shale rather than returning to the surface. That is to say, it is sequestered in the shale; just the thing we need to minimize releases of CO_2 into the atmosphere.

On the other hand, it costs more than water, and there is little information on its ultimate fate and on how to separate the volume of it that does return to the surface from the hydrocarbons.

The preliminary results of the study, all carried out using computer models based on known physical principles, are encouraging. Particularly encouraging is an effect seen in the sCO_2 and not in water: an abrupt cooling of about 200°C when the pressure is initially released, further shocking and potentially further enhancing crack propagation.

So, here we have a potential win-win situation; an economic reason to inject CO_2 under pressure into shale fields where it may be sequestered, and an increase in the effectiveness of oil and gas production by fracking with potentially less impact than the current hydraulic practice. Time will tell.

Middleton, R., Viswanathan, H., Currier, R., Gupta, R., 2014. CO_2 as a fracturing fluid: Potential for commercial-scale shale gas production and CO_2 sequestration. Energy Procedia 63, 7780-7784.

Heavy Oil Production using Fracking and Microwaves
by Emil Morhardt

Fracking isn't just for natural gas and conventional oil; it also increases the production of heavy oil in low permeability reservoirs, but if the oil is heavy enough and the cracks don't penetrate very far, the flow rates decay rapidly and not much oil is recovered. Heavy oil, not defined in this paper by Davletbaev *et al.* (2014), is usually a term used for oil just a little less viscous than bitumen, the "extra heavy oil" found in the highly contested Canadian tar sands—more or less like asphalt. One possibility is to heat up the rock surrounding the well to make the oil less viscous. A technique that works in conventional oil wells is to inject steam, but with heavy oil in low permeability reservoirs steam doesn't work very well. The approach discussed in this paper is to use microwave radiation (also known as radio-frequency electromagnetic radiation) of the sort used in your kitchen microwave oven. Of course it is impossible to bring the oil-shale to the oven—the oven has to go to it in the form of downhole electrodes. Another difference is the amount of energy required. A "powerful" home microwave oven consumes about 1,000 Watts (1 kW) provided by a standard kitchen electrical circuit. The heating of well bores simulated in this paper used 10–30 times that much electricity, but experimental studies have shown that after a day-and-a-half of heating, temperatures in the well can exceed 300 °C (572 °F) and can raise the temperature of the shale (and oil) to over the boiling point of water a few meters away.

This paper is about an operational strategy to maximize the flow of heavy oil using a combination of fracking and microwave heating, while minimizing the amount of electricity required. The idea is to frack first and let as much oil out as will come, then to heat for a while, stopping when the oil starts flowing, and reheating as many times as it takes to get as much of the oil as economically possible. The study compared the simulated oil production from a "cold" fracked well with that from a heated one for 550 days, assuming that the microwave device cost $100,000 and the oil sold for $100 per barrel. The authors concluded that the multi-stage heating could increase oil production by 87% in a well with low permeability "short" fractures, the type of well most suitable for this technique. Depending on the permeability of the fractures, the oil production rate, the price of oil, the amount of electricity used, and few other variables, payback times for the microwave heating ranged from 420 days to five-and-a-half years; somewhat less than the 550 days the simulations were run to four times that. The authors don't comment on what this says about the feasibility of using microwaves to increase heavy oil production, but from my calculations, it doesn't look like a very good option for most of the cases they tested.

Davletbaev, A., Kovaleva, L.A., Babadagli, T., 2014. Heavy Oil Production by Electromagnetic Heating in Hydraulically Fractured Wells. Energy & Fuels. Just Accepted Manuscript DOI: 10.1021/ef5014264.

Fracking: Fix it or Forget It? Global Gas and Oil Prices Falling

by Emil Morhardt

Daniel E. Klein, an energy industry consultant, writes an interesting piece about fracking problems in Natural Gas & Electricity, an industry newsletter. His approach is to look at the prognostications of the Energy Information Administration Annual Energy Outlook (AEO)—pretty much the bible of energy projections—as they have changed from 2000 to projections of where we will stand in 2040. For example, there wasn't much shale gas until 2005 and in 2005 the AEO predicted that US natural gas imports would increase sharply in the near future. The 2014 projection, however shows the opposite: a steady increase in US exports, at least through 2024. Similarly, "peak oil" in the US has also been reversed by shale oil production, with the crude oil production in 2013 the highest in 25 years, and imports falling sharply, at least so far. OPEC has been debating, on the one hand, decreasing oil production, so as to increase global oil prices and therefore revenues or letting production stand so as to lower prices even further to put price pressure on American fracking operations. For now they have settled on the latter, but in the short-term oil prices will have little effect on American oil operations.

A few years ago the AEO projected that natural gas would only ever constitute 10–20% of fuel for power generation; now it seems to be displacing coal sufficiently to account for 25% now, and 30% by 2040. This has led not to "peak oil", but to "peak CO_2". It looks as if we are on a permanent CO_2 emission downslope in the US without ever having had to bite any bullets; we are leading the world in CO_2 reduction purely because the price of shale gas in the US has made it competitive with coal. When you hear the politicians from coal-producing states blaming the Obama administration for their troubles, it is therefore largely disingenuous.

The bottom line of this article is that whatever the environmental problems with fracking, they can be fixed, and the benefits are so monumental that they ought to be. I imagine that this is more-or-less correct, but there's probably nothing as disconcerting as having the bucolic farmland that once surrounded you turned into a heavily industrialized oil field— that's almost impossible to fix. However, unless one is constitutionally opposed to fracking, one ought to hope that it can replace coal globally with time...at least long enough for renewables to come of age.

Klein, D.E., 2014. Fracking: Fix It or Forget It? Natural Gas & Electricity 31, 1-8.

Fracking in South Texas: Spatial Landscape Impacts
by Emil Morhardt

In a Master's thesis from the University of Texas at Austin, Jon Paul Pierre presents an interesting analysis of the effects of development (which includes a good deal of horizontal drilling and hydraulic fracturing) in the Eagle Ford Shale play in South Texas, where more than 5,000 wells have been drilled since 2008. What he sets out to do is assess the spatial fragmentation of the landscape from the construction of drilling pads, roads, pipelines, and other infrastructure. He used 2012 aerial photography with a 1-m resolution obtained from the National Agricultural Imagery Program (NAIP), and over laid on that the locations of well pads, pipelines, and other infrastructure, then used Geographical Information System (GIS) tools to characterize the types of areas being disturbed.

For the 628 wells in La Salle County (a portion of the overall study area) for which there was evidence of associated infrastructure on the aerial imagery, pipeline disturbance occupied 97 square kilometers, and drilling pads, 17 square kilometers. These activities heavily disturbed 3% of the county area, but 8.7% of the core areas, with a reduction in overall vegetated area from 91% to 89%, and of forest area from 76% to 68%. Probably of more concern than the absolute loss was the ecological spatial fragmentation caused mainly by the pipelines, potential soil loss from wind erosion of the disturbed areas, and interference with normal drainage patterns.

There aren't any particularly novel conclusions from this research, but it is a good example of how to analyze the physical effect of large-scale well development (or any kind of spread-out industrial development) on the landscape. Nevertheless the landscape is pretty thoroughly disrupted by oil and gas development, maybe even more so than this thesis implies.

Pierre, J.P., 2014. Impacts from above-ground activities in the Eagle Ford Shale play on landscapes and hydrologic flows, La Salle County, Texas. University of Texas at Austin, Master of Science Thesis, August 2014.

Try Not to Live Too Close to a Fracked Well
by Emil Morhardt

If you happen to live within 1 km of a hydraulically fractured well in Pennsylvania, and you get your water supply from a well, you stand about twice as large a chance of having skin and upper respiratory problems than if you live 2 km or farther away; you have over 3 health symptoms, on average—people further away have only 1.6. Looked at another way, 13% of people living near fracking operations have upper respiratory problems, versus 3% living farther away; and 39% of the same group of people have upper respiratory problems versus 18% living further away. That is the disturbing result of an epidemiological study of almost 500 people in an area of natural gas drilling in the Marcellus Shale, just published in a journal of the National Institute of Environmental Health Sciences (Rabinowitz *et al.* 2014).

A major limitation of the study is that these conditions were all self-reported, and it is within the realm of possibility that people living near gas wells were more sensitive to the possibility of health problems, and more likely to report less severe ones. On the other hand, people living close to the wells might be receiving revenue from them and be more hesitant to report health problems...nobody knows yet. Clearly there needs to be a follow-up study in which health professionals characterize the symptoms.

What caused the reported symptoms? It could be contaminants leaking into the water or air from the fracking operations; or it could be related to the increased stress and anxiety of living near them. This study raises more questions than it answers, but that's the way it is with science, and it certainly adds to the body of knowledge suggesting that fracking is not without serious issues and in need of better regulation.

Rabinowitz, P.M., Slizovskiy, I.B., Lamers, V., Trufan, S.J., Holford, T.R., Dziura, J.D., Peduzzi, P.N., Kane, M.J., Reif, J.S., Weiss, T.R., 2014. Proximity to Natural Gas Wells and Reported Health Status: Results of a Household Survey in Washington County, Pennsylvania. Environ Health Perspect.

Biotic Impacts of Fracking
by Emil Morhardt

How does shale-bed energy development, including hydraulic fracturing, affect ecology? There have been a number of studies looking into this, and a new review paper by Sara Souther at the University of Wisconsin and seven colleagues at a diverse array of other institutions summarizes the current knowledge and where the gaps are in it. Their legitimate fear is that substantial damage will be done before much is known about the issues, and there is plenty of experience with other rapid industrial development to warrant concern. As an example, consider the damming of nearly all the rivers on both the east and west coasts of the US with little attention paid to the consequences for salmon.

The big issues they identify are subsurface and surface water contamination by fracking fluids, diminished stream flow because of water diversions for fracking, habitat loss and fragmentation, general disturbance to wildlife from the noise, light, and air pollution of fracking operations, and, of course, the atmospheric increase in greenhouse gases resulting from both leakage of natural gas in the process of collecting it, and CO_2 when the gas is burned by end users.

As the authors correctly point out, most of these are difficult to study, not much studied yet, and not evidently being studied to the degree necessary to properly evaluate the impacts.

They'd like to see that change, since it doesn't appear that oil shale development is likely to slow down any time soon.

Souther, S., Tingley, M.W., Popescu, V.D., Ryan, M.E., Hayman, D.T., Graves, T.A., Hartl, B., Terrell, K., 2014. Biotic impacts of energy development from shale: research priorities and knowledge gaps. Frontiers in Ecology and the Environment 12, 330-338

Methane Emissions in Colorado Exceed EPA Estimates; Fracking?

by Emil Morhardt

Colorado's north Front Range, north of Denver and east of Boulder and Fort Collins has become a frackers' paradise, with 24,000 active wells in 2012, 10,000 of them drilled since 2005. In the hot muggy summers, volatile organic compounds, including methane, ethane, propane, butane, pentane, and sometimes the carcinogen, benzene,(all commonly found in oil and natural gas (O&G), accumulate in the air, leading to elevated ozone levels, and contributing to global warming. Previous estimates of the total amounts released were based on a combination of bottom-up estimates of releases from various sources based on a variety of sampling methods, as well as air samples from tower sampling stations. Extrapolating these to the whole O&G area carries all of the uncertainty associated with each of these estimates. In order to get a top-down, fully integrated estimate, Pétron *et al.*, research scientists at NOAA, sampled the area from an airplane equipped with an instrument that continuously recorded methane, carbon dioxide, and carbon monoxide concentrations, and was also capable of taking discreet air samples for measuring other volatile organic compounds typically released from O&G operations. They found that the concentrations of most volatile organic compounds were twice as high, and that of benzene was seven times as high as previously estimated by the state of Colorado, and the hourly emissions rate was three times as high as estimated by the USEPA. The bottom line is that a lot more methane and other volatile organic carbons are being released from the O&G operations than was previously thought.

The researchers made a systematic effort to factor out the non-O&G methane emissions from the many agricultural sources in the region, using county-level agricultural census statistics from Colorado and the USEPA to determine the head count for several categories of cattle, as well as for sheep and poultry. This they multiplied by standard methane emission factors for each type, assuming a 20% uncertainty in both head count and emission factors—that their estimated head count was within 10% of the permitted capacity supported its accuracy. They also estimated emissions from manure management, but with less certainty owing to a lack of information on the details of the individual management systems. Additionally they estimated releases from the municipal landfills and both municipal and industrial wastewater treatment plants. All of this resulted in a "bottom-up" estimate of methane emissions from all non-O&G sources, which allowed them to attribute the rest to the O&G operations. The aircraft also took flask samples to measure non-methane hydrocarbon concentrations, and all of these were linearly correlated with the methane concentrations, suggesting the same source, presumably O&G operations.

It isn't clear why there is such a large discrepancy between bottom-up and top-down estimates of O&G-related volatile organic compound releases, but a reader of this paper would infer that there is not nearly enough monitoring of ground-based releases to give any sort of accurate overall results. There are as many as several thousand new wells being drilled and hydraulically fractured each year in their study area but, according to the authors, only a subset of them is subject to inspection; if they are drilled by the largest of the over 100 companies in the area that subset is inspected every 3 years; if they belong to a smaller company, only every 5 years. Clearly there is inadequate ground-based monitoring going on if a comprehensive view of O&G releases is desired.

Are any of these releases attributable to fracking? Virtually all of the new wells are fracked, but this study design had no way of determining at what stage of the production cycle the releases occurred.

Pétron, G., Karion, A., Sweeney, C., Miller, B.R., Montzka, S.A., Frost, G.J., Trainer, M., Tans, P., Andrews, A., Kofler, J., 2014. A new look at methane and nonmethane hydrocarbon emissions from oil and natural gas operations in the Colorado Denver-Julesburg Basin. Journal of Geophysical Research: Atmospheres.

Unexpectedly High Methane Concentrations over Pennsylvania Shale Gas Fields Too
by Emil Morhardt

Methane, the main constituent of natural gas (both that from gas wells and from farm operations) is a powerful greenhouse gas, around 30 times more potent than CO_2 over the hundred years after it is emitted. It is on the rise in the air above Pennsylvania, and the culprit might be shale gas development, which utilizes hydraulic fracturing. Caulton *et al.* (2014) used an airplane to sample the air above a 2,800-square-kilometer area of the Marcellus shale formation gas fields in Pennsylvania. It was rich in methane, with between 2 and 15 grams heading skyward over each square kilometer every second, the upper limit of which is quite a lot higher than the 5 grams estimated from what was previously known about wellhead methane emissions; the authors suspected that the transient nature of gas leakage might be the reason, making very difficult to come up with an average over time from ground-level measurements. Since they were in an airplane, however, they could circle around areas of high concentrations and pinpoint the source. It turns out that, sure enough, the sources were well pads and, in one case, a coal mine, but the interesting thing is that these wells were in an early state of development, hadn't reached their full depth, and hadn't yet been hydraulically fractured. The large releases at this stage of development were completely unexpected, but the culprit pads were visibly missing commonly used approaches that might have prevented the releases (shale shakers, mud pits), so it may have been carelessness on the part of the drillers causing the releases.

Similarly high measurements have been obtained from flights over well fields in Texas, Oklahoma, and Kansas, suggesting a widespread national problem. At this stage it doesn't look like the problem is fracking, so much as failure to employ standard techniques to capture the methane and save it or flare it, but clearly something needs to be done.

Caulton, D.R., Shepson, P.B., Santoro, R.L., Sparks, J.P., Howarth, R.W., Ingraffea, A.R., Cambaliza, M.O., Sweeney, C., Karion, A., Davis, K.J., 2014. Toward a better understanding and quantification of methane emissions from shale gas development. Proceedings of the National Academy of Sciences 111, 6237-6242.

How Long will the Fayetteville Fracking Play Last?

by Emil Morhardt

How long will shale gas be available until it plays out? The Bureau of Economic Geology (BEG) at the University of Texas at Austin is making a concerted effort to find out for the four largest shale plays currently in development in the US. The first they reported on was the Barnett Shale in Texas. The topic of this post is their second study, conducted on the Fayetteville (Arkansas) Shale by John Browning and eleven colleagues at the BEG. The overall answer is a long time—but well short of a century—with production peaking soon and falling to between half and a third of the current levels by 2030 and continuing to decline thereafter; they ran their model through 2050 and estimate the technically recoverable gas resources if economics were not an issue (38 trillion cubic feet), and the amount likely to be recovered eventually given economic reality, about half that.

The article. Published in the Oil & Gas Journal (Browning *et al.* 2014) gives insight into the complex modeling required to come up with a credible answer. The authors looked at the gas production history of every (3,689) well drilled in the basin up to 2011, mapped their estimated 30–year productivity in a fine grid across the 2,737 square mile study area, extrapolated that productivity to nearby areas, and concluded that only about a third of the economically feasible wells have yet been drilled. As one would expect, all of their estimates are highly dependent on assumptions entered into the model, perhaps the most uncertain of which is the price of natural gas—if it is high, many more wells will be drilled than if it is low, but the model can accommodate just about any likely variation, so as more information appears over time, it will be a snap to rerun it.

I recommend this paper for its clear treatment of a difficult subject. For an example of a much more technically challenging description, try reading Patzek *et al.* (2013) by three of these same authors.

Browning, J., Tinker, S.W., Ikonnikova, S., Gülen, G., Potter, E., Fu, Q., Smye, K., Horvath, S., Patzek, T., Male, F., 2014. Study develops Fayetteville Shale reserves, production forecast. OIL & GAS JOURNAL 112, 64-+.

Iceland's Turning Greenhouse Gases Into Stone

by Hannah Brown

Positioned near the Hellisheidi Power Plant in Iceland, researchers at CarbFix, a $10 million project funded by Reykjavic Energy, the United States Department of Energy, and the EU, among others, combines water and carbon dioxide, compressed to the point that is in its liquid form, and injects the mixture thousands of feet down into balsatic rock, a reactive volcanic rock that makes up almost the entirety of Iceland's foundation, as well as the Mid-Atlantic Ridge in general. The combination of carbon dioxide and water interacts with the rock as it releases calcium and magnesium and turns into the original mixture into limestone. Initially the model predicted that the process would take 5 years but CarbFix has found that it happens much faster than expected, essentially completing the transformation of carbon dioxide into limestone within one year (or.is).

The team involved prefers not to label this method as a form of CO_2 storage because, in fact, the CO_2 is completely transformed. Instead, they refer to this process as "mineral carbonation." The researchers say that this alternative to CO_2 storage is, though initially more costly, cost-effective as the years progress. They argue that as opposed to CO_2 storage, there is no chance that the CO_2 could leak during its transformation process and therefore does not need the extensive monitoring that storage options require.

While this is all intriguing, some observers of CarbFix see the project as a misuse of funds and a misdirection of thought. Instead of funding research that releases us from the grasp of fossil fuel dependency, CarbFix and its procedure act almost like a bandaid. Instead of addressing the crux of the problem, they pose a superficial fix to the issue's consequences. In this pessimistic view, CarbFix is unnecessary, but I believe that just as Dr. Matter, one of the lead CarbFix researchers and geochemists at the University of Southampton, says in an interview with The New York Times "the problem is big enough...we need many solutions." CarbFix is not going to solve the issue of climate change on its own, but it can have an impact and ameliorate the negative effects that are currently occurring (nytimes.com).

Fountain, Henri. "Turning Carbon Dioxide Into Rock, and Burying It" NYTimes.com. February 9, 2015

Halperin, Carrie and Fountain, Henry. "Fixing Climate in Iceland" NYTimes.com. February 9,2015

Reykyjavic Energy. "CarbFix" Orkuveita Reykavikur. or.is

Underground Storage of CO_2: Attempts to Eliminate Carbon Emissions

by Nour Bundogji

Postdoctoral researchers, Yossi Cohen and Daniel Rothman, at Massachusetts Institute of Technology, recently published an article in the Royal Society Proceedings on the effectiveness of storing carbon dioxide underground in an effort to decrease carbon emissions in our atmosphere. When I first read this I immediately envisioned suction cups elevated high into earth's atmosphere connected to long pipes extended deep within earth's crust. Yet, you guessed it, the technology is quite different. Instead, greenhouse gases emitted by coal-fired power plants would be pumped into salt caverns 7,000 feet underground where these gases would react with the salt water and solidify (Cohen and Rothman, 2015). The US Environmental Protection agency estimated that this technology could eliminate up to 90% of carbon emissions from coal-fired facilities. Considering the current state of our ozone layer and the drastic climate changes we've been experiencing these past years, this seems like a promising step forward in saving our environment. However, commentators on this technology, like Christopher Martin from Bloomberg, pointed out a few flaws. I knew it was too good to be true.

Cohen and Rothman expected that much of the carbon dioxide would solidify. However, upon micro and macro-scale analysis and numerical simulation, they noted that solidification of these greenhouse gases only occurs at the surface of these salt caverns creating a barrier that prevents more gas from reaching the brine. As Martin simply puts it, "it's clogging up the process" (Bloomberg.com). If these carbon gases do not reach these saltwater pockets, the carbon sources remain in the gaseous or liquid phase and can escape back into the atmosphere very easily, putting us back to square one with respect to controlling climate changes.

Although these clogs can occur, I don't feel we've hit a roadblock with this advanced storage process. Instead, I feel like this is a strong move in the right direction especially since significantly less carbon dioxide will precipitate with this technology.

Cohen, Y. and Rothman, D.H. 2015. Mechanisms for mechanical trapping of geologically sequestered carbon dioxide. Royal Society Proceedings A. Jan 21, 2015.
http://rspa.royalsocietypublishing.org/content/471/2175/20140853
Martin, Christopher. 2015. "Underground Carbon Dioxide Storage Process Faces Clogs, MIT Says." Bloomberg. Jan 21, 2015
http://www.bloomberg.com/news/2015-01-21/underground-carbon-dioxide-storage-process-faces-clogs-mit-says.html

ENERGY GOVERNANCE

A good bit of the energy needed to provide energy must be expended in governance. Often not enough is spent to achieve a desirable outcome. What follows are some thoughts on the institutional processes surrounding the supply of energy.

Nigerians Push for Renewable Energy to Solve Power Crisis
by Nour Bundogji

For many years now, Nigeria has been facing an extreme electricity shortage. Why? Well let's look at Nigeria by the numbers. Overall, Nigeria consumes 2.5 million barrels of oil a day. Additionally, Nigerians spend, on average, ten dollars a week for a grid of electricity regardless of whether their grid works or not. Lastly, Nigeria spends $5 billion a year on fuel to generate electricity leaving about two thirds of Nigerians with no access to electricity (which is more people without electricity than any other country in the world except for India). Kennedy-Darling and her colleagues at University of Chicago reason that these energy deficits are a result of financial and structural problems in Nigeria's current energy system. These problems demonstrate a ripple effect where the decreased efficiency of the energy producing capacity in Nigeria (a structural problem) leads to low productivity, excessive debts, and high fixed costs associated with power (financial problems) (Mohammad, 2007).

To put this into perspective, modern technology, which Nigeria does not have access to yet, allows for 40% of the energy consumed in thermal plants to be converted to electrical energy. However, in Nigeria only 12% of energy is converted (Sambo, 2005).

In a recent article posted on DW, Damon van der Linde and Johan Demarle noted that some residents are now seeking alternative solutions to the country's ongoing crisis. For instance, these past few years there have been a series of expos on alternative energy in an effort to explore the possibilities of renewable energy in Nigeria. As a result of these expos, foreign companies like Lumentech, a solar panel company form South Africa, took interest in Nigeria as possible country for investment. Pieter Joubert, the production manager at Lumentech, is looking for ways to expand into other African countries and he said, "It's a very, very big market and that is why we're expanding and hopefully we can improve people's lives right through Africa." With these foreign invests Nigerians are now able to take advantage of what renewable energy has to offer. One Nigerian shopkeeper purchased a $50 solar panel from Lumentech claiming that he is "no longer spending any money on energy, but instead on what to cook." This got many of his fellow neighbors interested increasing the demand for more renewable energy sources. Thus, the Nigerian government started a renewable energy program that would allow its citizens access to renewable energy sources like solar panels. Unfortunately, there has been some opposition from local investors toward funding this program since "those with the financial capability to do so have invested a lot of money on generators...so renewable energy is a threat to their business" says Larry Edeh the organizer of The Nigeria Alternative Energy Expo. Nonetheless, foreign investments from renewable energy companies have helped in decreasing this power crisis in Nigeria. I just hope that local investors see promise in this cause soon since the need for sustainable and reliable energy will only increase.

Demarle J. and Linde D. 2015. Nigerians turn to renewable energy as solution to power crisis. European Journalism Centre. Jan 27, 2015.
http://www.dw.de/nigerians-turn-to-renewable-energy-as-solution-to-power-crisis/a-18216818
Sambo, A. S. "Renewable Energy for Rural Development: The Nigerian Perspective". ISESCO Science and Technology Vision 1 (May 2005): 12-22.
Mohammed, K. "Why Electricity in Country Is in Unhappy Condition (1)". Africa News. 3 December 2007. Lexis Nexis Academic. <web.lexis- nexis.com/universe>
Kennedy-Darling, J. Hoyt, N. Murao, K. Ross, A. The Energy Crisis of Nigeria An Overview and Implications for the Future. The University of Chicago. June 2008.
http://franke.uchicago.edu/bigproblems/Energy/BP-Energy-Nigeria.pdf

The New European Energy Union Faces Some Critiques
by Nour Bundogji

The World Nuclear News recently discussed The European Commission's intention to begin a new European Energy Union. This initiative will help "reform how Europe (EU) produces, transports, and consumes energy."

The meeting took place on February 4, 2015 in Riga, Latvia where The European Commission outlined the intention and goals of this initiative. It's primary aims are: "diversifying energy sources currently available to Member States, helping European countries become less dependent on energy imports, making the EU number one in renewable energy in the world, and leading the fight against global warming." With these aims, the European Commission listed five goals for the European Energy Union, which are:

1) ensuring security of supply
2) building a single internal energy market
3) raising energy efficiency
4) decarbonizing national economies, and
5) promoting research and innovation.

Sounds appealing, at first. However, Forbes contributor, James Conca, points out a few nuances within this initiative making this Energy Union a little more difficult to implement than initially expected.

For instance, Conca interestingly pointed out that the second goal (building a single internal energy market) would require much enforcement while the fourth goal (decarbonizing everyone's economies) "might not sit well if you have a definition for decarbonizing that differs from Germany's (EU report)."

These goals also present a few unanswered questions like: "Will England insist on more wind in Ireland, against their will? Will Germany push for greater dependence on Russian natural gas? Will Germany try to shut down France's nukes, even though they produce more carbon-free electricity than all other low-carbon sources in Europe combined?"

Nonetheless, the commission urges that the time is right for the creation of this Energy Union since "energy security is high on the political agenda, and a door for an ambitious climate agreement in Paris at the end of 2015 was opened in the European Council last October." Furthermore, "the recently adopted Investment Plan for Europe is designed to unlock the financial means the energy sector really needs. The currently low oil prices are also giving an extra incentive and give more political and financial room to do what is necessary to achieve a more competitive, secure and sustainable European energy policy." However, Conca claims that these "low oil prices are only the fallout of a battle between Saudi Arabia and the United States to bankrupt the American oil shale industry. And that can't last too long before one side breaks."

Maria van der Hoeven, the Executive Director of the International Energy Agency, was also at the Riga conference.

She noted, "In the coming decades the EU is expected to retire half of its electricity capacity. Nuclear plants are ageing, with half of the EU's existing nuclear capacity to be retired by 2040. Environmental rules also require the phase-out of old coal fired power plants." On the other hand, it seems that "this concept doesn't sit too well with Europe's largest companies," states Conca. For instance, CEOs of European energy companies like Germany's E.ON, France's GDF Suez, and Italy's Eni disagree with the decarbonization plan and have blatantly stated that the stability of Europe's electricity generation is at risk from the warped market structure caused by skyrocketing renewable energy subsidies that have swarmed across the continent over the last decade (Géraldine Amiel WSJ). Furthermore, there will be some push-back from Europe's industries since renewables and their subsidies are a major component of the new European Energy Union's strategy— a conflict of interest for these corporations (Capgemini).

Lastly, Vice president of the Energy Union, Maroš Šefčovič, said, "We will work to ensure a coherent approach to energy across different policy areas, to create more predictability."

Sounds great!

But, as Conca notes, "there is the problem of integrating the diverse regulatory and market trading systems that would have to become congruent across all borders of Europe." For example, some EU countries want to exploit domestic fossil fuels (Germany's expansion of dirty coal) while others stress the need for lower carbon sources (UK's increased use of natural gas).

All in all, the European Energy Union is scheduled to release later this month, on February 25th. However, the commission needs to move fast to accommodate for these subtleties if they want the Energy Union to succeed.

Conca, J. The Role of the New European Energy Union. Forbes.com. Feb. 9, 2015.
http://www.forbes.com/sites/jamesconca/2015/02/09/the-role-of-the-new-european-energy-union/
World Nuclear News. Europe Starts Work on Energy Union. Feb. 6, 2015.
http://www.world-nuclear-news.org/EE-Europe-starts-work-on-Energy-Union-0602154.html
Capegemini. European Energy Markets Observatory 2013. Oct. 10, 2013.
http://www.capgemini.com/resources/european-energy-markets-observatory-2013-full-study
Ameil, G. Energy Bosses Call for End to Subsidies for Wind, Solar Power. Oct. 11, 2013.
http://stream.wsj.com/story/latest-headlines/SS-2-63399/SS-2-352276/

National Parks or Energy: Kenya's Dilemma

by Jessie Capper

According to a recent report released by the International Energy Agency's 'Africa Energy Outlook,' unreliable power supply has been a persistent problem in African countries. The IEA claims that by addressing this uncertainty, African governments help increase investment in their respective country's power sector, and ultimately boost their GDP by an estimated $15 (International Energy Association 2014). Kenya continues to address its problems with efficient, reliable, and high-cost energy through the pursuit of renewable energy sources—varying from solar and wind power, to hydropower, and geothermal energy. Although Kenya's energy initiatives are progressive and admirable, there is rising concern over detrimental side effects, especially to the national parks.

Geothermal power, in particular, has been a prominent energy source for Kenya (Out-Law Feb 25, 2015). Many international organizations, including the World Bank, Japan International Cooperation Agency, and the European Investment Bank, sponsored the Olkaria geothermal plant in Kenya's Rift Valley. According to the World Bank, geothermal power accounted for only 13% of Kenya's overall energy in 2010, compared to the estimated 51% that it now provides for Kenya's energy grid (Out-Law Feb 25, 2015). The Kenya Electricity Generation Company (KenGen) claimed that geothermal energy is among the cheapest and most reliable renewable sources for electricity in Kenya as it costs a mere 7.2 US cents per kilowatt hour; furthermore, geothermal power is not dependent on weather patterns as other renewable sources are, which ultimately establish unreliable energy supplies for Kenyans (Out-Law Feb 25, 2015).

The Olkaria geothermal plant in Kenya's Rift Valley is a rich source of power for Kenya's electrical grid (Olkaria Geothermal Project). The Olkaria Project produces enough energy to serve at least 500,000 Kenyan homes with 280 mW of geothermal power, and there is currently discussion of expanding this to 5,000 geothermal mW by 2030 (Sierra Leone Times Feb 11, 2015). The site has proven to reduce power prices, and is estimated to increase Kenya's economic growth by substantially expanding beyond the mere 23% of Kenyans who currently have access to electricity (Sierra Leone Times Feb 11, 2015). Unfortunately, this development presents damaging consequences for Kenya's national parks. The Olkaria Geothermal facility is set inside Hell's Gate—a national park in Kenya (Sierra Leone Times Feb

11, 2015). Expanding the facility will undoubtedly lead to a disruption in Hell's Gate ecosystem according to conservationist Silas Wanjala. Wanjala claims that the vegetation and animal life have changed due to the power generation in the park; animal species, such as the buffalo and white-backed vulture, are not nearly as healthy or viable as they once were before the existence of the Olkaria project (Sierra Leone Times Feb 11, 2015).

Kenya's dilemma with the Olakria Project exemplifies the conflict facing numerous African governments—how to properly meet a country's increasing demand for electricity while conserving the wildlife within and beyond its national parks. The Kenyan government is hopeful that it can develop a plan that ultimately achieves both goals in an effective and efficient way. Hopefully, their solution will establish a replicable model for other developing countries facing the same issue.

"Africa Energy Outlook." International Energy Agency. 2014. Accessed February 27, 2015.
    http://www.iea.org/publications/freepublications/publication/WEO2014_Africa EnergyOutlook.pdf
"Kenya's Green Energy Boom Could Threaten National Parks." Sierra Leone Times. February 11, 2015. Accessed February 25, 2015.
    http://www.sierraleonetimes.com/index.php/sid/230133683
Olkaria Geothermal Project (http://www.worldbank.org/projects/P001287/olkaria-geothermal-power-expansion-project?lang=en)
"Power Plant Investment in Kenya Boosts Geothermal Production to 51%." Out-Law. February 25, 2015. Accessed February 26, 2015. http://www.out-law.com/en/articles/2015/february/power-plant-investment-in-kenya-boosts-geothermal-production-to-51---/

A Call for Holistic and Immediate Action: The World's Peaking Renewable Resources
by Jessie Capper

Petroleum is universally accepted as a limited resource, and everyone knows it could all be used up, but according to landscape ecologist Ralf Seppelt, many people incorrectly assume that biological sources are "infinitely available because all plants need are water and sunshine" (Hart Feb 2, 2015). Unfortunately, that is not true. According to scientists at the Food and Agricultural Organization of the U.N., Michigan State University's Center for Systems Integration and Sustainability, and the Helmholtz Centre for Environmental Research—18 of the 20 biological resources they studied reached their peak in production between 1988 and 2008, including: meat production, global fish catch, milk, and soybeans (Hart Feb 2, 2015).

It is difficult to understand why so many renewable resources reached their peak during roughly the same time, but Seppelt speculates that expanding populations and changing diets in India and China may be responsible—primarily for renewables used for nutrition and feeding animals (Hart Feb 2, 2015). Additionally, resources tend to be interdependent; for example, producing food requires land, energy, and water, which often also creates pollution, aggravating the situation further (Science 2.0. Jan 14, 2015). Another possible explanation is that the availability of remaining resources is diminishing due to the process of extracting less accessible resources; researchers at MSU's Center for Systems Integration and Sustainability demonstrate that removing these resources has resulted in escalated ecological and economic costs for every unit extracted (Science 2.0. Jan 14, 2015).

According to Jianguo Liu at MSU's Center for Systems Integration and Sustainability, a holistic approach rather than a "one for one" substitution method is the most appropriate strategy to address these sustainability problems (Science 2.0. Jan 14, 2015). Liu claims that this issue of renewables reaching their peaks is "a call to not be simplistic when seeking solutions and to acknowledge that all resources come with costs, and that the costs may not be immediately obvious, and span the globe" (Science 2.0. Jan 14, 2015). Multiple resources are dissipating within a very narrow range of each other; therefore, combating this issue most effectively requires a universal and immediate strategy.

Hart, Tom. "World's Renewable Resources Reach Limit." Geographical. February 2, 2015. Accessed March 7, 2015. http://geographical.co.uk/people/development/item/756-world-s-renewable-resources-reach-limit.

"Sustainability Crisis: Renewable Resources Growth Has Been Maxed Out Since 2006." Science 2.0. January 14, 2015. Accessed March 7, 2015. http://www.science20.com/news_articles/sustainability_crisis_renewable_resources_growth_has_been_maxed_out_since_2006-152279.

Can California Continue to Lead in the Renewable Energy Industry?
by Jessie Capper

California is currently leading the renewable energy industry with Governor Jerry Brown's progressive 2011 bill requiring utilities to receive 33% of their power from renewable energy sources. In January, Governor Brown reported that California was "on track" to fulfill its 2020 goal of 33% renewable portfolio standard (RPS), which he hopes to increase to 50% by 2030 (Cheeseman Mar 16, 2015). Although multiple renewable energy projects have been completed and are outlined for upcoming years throughout the state of California, the local, state, and federal government are facing a few obstacles in the planning and implementation process of projects in Southern California.

The state and federal governments recently announced that they would proceed on fewer than half the California desert land originally designated for renewable energy projects. Regulators have currently agreed to develop on approximately 10 million acres of federal land out of the 22.5 million acres initially intended for solar, wind, and geothermal projects (Spagat Mar 10, 2015). The 12.5 million remaining acres are privately owned and require approval from local governments prior to construction of any renewable projects. Andy Horne, Imperial County's deputy executive on renewable energy plants, claims that some residents are concerned that these renewable energy plants will replace farming jobs in one of the nation's highest unemployment regions. Others involved in the projects are worried about the potential environmental damage associated with these renewable energy plants (Spagat Mar 10, 2015). Additionally, the 10 million acres under federal management is home to a diverse range of wildlife and ecosystems and are believed by some to be potentially off-limits due to conservation issues (Spagat Mar 10, 2015).

As the decision on how to move forward is currently at a stalemate, it is evident that the local, state, and federal governments must be involved at every point of the discussion process. According to Laura Crane, associate director of The Nature Conservancy's California land program, "Given the size and complexity, it's understandable why the agencies are breaking it apart in pieces they can manage... If the counties aren't happy—and they clearly weren't—it's really important that it be slowed down. If the counties don't agree, it can't be implemented" (Spagat Mar 10, 2015).

Cheeseman, Gina-Marie. "California Leads the Nation In Solar Power." Triple Pundit People Planet Profit. March 16, 2015. Accessed March 25, 2015.
http://www.triplepundit.com/2015/03/california-leads-nation-solar-power/.
Spagat, Elliot. "US, California Modify Area for Renewable Energy Plants." Press-Telegram. March 10, 2015. Accessed March 25, 2015.
http://www.presstelegram.com/business/20150310/us-california-modify-area-for-renewable-energy-plants

Psychology's Role in Being Green
by Alex Elder

Although global climate change is a well-documented phenomenon with severe implications for the future of our planet, environmental issues are often paid little attention by politicians and laypeople alike. Other issues like the economy or job security are viewed as top-priority concerns, even though continued damage to the environment could potentially endanger the human race. Small changes in our everyday lives could make a significant positive impact, but many people are not concerned enough about the environment to make a change.

But why are environmental issues perceived as less important than other, less detrimental problems? Elke Weber, a psychologist at Columbia University, seeks to answer this question through her research on risk assessment and how people perceive climate change. She is an expert on the psychological components of judgment and decision-making in the face of uncertainty, specifically in environmental contexts. She founded the Center for Research on Environmental Decisions (CRED), which investigates human adaptation to climate change and how to enable further sustainability efforts. Through this organization, Weber published an influential paper which found that people view time-delayed, abstract risks as less serious threats than more immediate, concrete risks, regardless of the severity of the outcomes (Weber, 2006). Thus, although the overall negative impact of climate change and other environmental problems is much greater than, say, possible unemployment, the former is viewed as less hazardous at the individual level.

But now that research has demonstrated this interaction between psychology and environmental issues, what can we do about it? Fortunately, by taking advantage of this new research, we can begin to represent environmental issues in a way that people will take more seriously. Once environmental initiatives are presented in a way that appeals to human patterns of decision making and risk assessment, then people will be more likely to become involved (Gertner, 2009). Elke Weber collaborated on a Climate Change Communications Guide that includes strategies on how to increase environmental engagement and which communications tactics to avoid. Based on the findings in Elke Weber's research, simply altering the presentation of information related to climate change could result in more sustainability engagement and a prioritization of the environment. When risks are perceived as more eminent and personal and less distant and abstract, people are more likely to prioritize that risk and the actions needed to avoid it. Thus, by appealing to psychological constructs involved in decision making, participation in environmental initiatives can be increased.

Weber, E. U. (2006). Experience-based and description-based perceptions of long-term risk: Why global warming does not scare us (yet). Climatic Change,77(1-2), 103-120. Connectingonclimate.org

Swim, J., Clayton, S., Doherty, T., Gifford, R., Howard, G., Reser, J., Stern, P., & Weber, E. (2009). Psychology and global climate change: Addressing a multi-faceted phenomenon and set of challenges. A report by the American Psychological Association's task force on the interface between psychology and global climate change. American Psychological Association, Washington.

Gertner, Jon (2009). Why Isn't the Brain Green? The New York Times. http://www.nytimes.com/2009/04/19/magazine/19Science-t.html?pagewanted=all

Hamburg is an Industrial City Reborn with a Renewable Energy Economy
by Liza Farr

Increasing regulation of fossil fuels and pollution, and the shift of jobs from industrial to tech has left many industrial cities with struggling economies. In Germany, the industrial city of Hamburg has fought this trend and is now known as the center of renewable energy for the nation. This past October, HusumWind, one of the world's largest wind power conferences, was held in Hamburg (Hales, Oct 9 2014). There are already 5,000 wind industry employers in the city, and that number is expected to double with the expansion of offshore wind facilities (Hales, Oct 9 2014). Nearly all the leading international wind companies have offices in the region (Hales, Oct 9, 2014). Twenty five thousand people are already working in renewable energy in Hamburg, and experts predict this number will grow by 40% by 2015 (Renewable Energy Hamburg, October 2012). Nineteen hundred and eighty green tech companies with 33,400 employees are based in the city (Hales Oct 9, 2012). The city is the central planning location for solar farms in Germany and across the world, and the most important development and management location for wind power in Germany (Renewable Energy Hamburg, October 2012).

Hamburg's success in transforming itself from an industrial port city to a hub for renewable energy is due to a number of concerted efforts by key players. The government has made the climate welcoming for these companies and put money into financing the projects. The mayor identified the renewable energy sector as one of the key areas for the city's prosperity, showing that the city government has incorporated the renewable energy sector as central to their economic planning (Hales, Oct 9 2014). The government also plans to make all their public transit buses zero emission vehicles within 20 years, further revealing the government's focus on renewable energy across departments (Montgomery, June 2 2014). The universities have also integrated renewable energy science and technology into their curriculum and research (Hales, Oct 9 2014). The Hamburg Metropolitan Region government is focused on developing the renewable energy sector, and they are supported by the Renewable Energies Cluster, a network of 185 member companies that jointly employ 25,000 staff (Maunder, March 24 2014). The cluster is meant to strengthen cooperation among companies, as well as providing an arena to pool research, skills, facilities, and institutions (Renewable Energy Hamburg, Oct 2012). Additionally, the region is naturally equipped with features helpful for renewable energies. The traditional crops in the region are excellent inputs for biomass and biogas production (Renewable Energy Hamburg, Oct 2012). The region also offers ideal conditions for wind and solar power (Renewable Energy Hamburg, Oct 2012). These conditions, along with the political support, business network, and reputation as European Green Capital in 2011 have formed Hamburg into a center for renewables in Germany, and even on the international scale. American rust belt cities could learn from Hamburg's success in transforming its economy using renewable energy. Even without ideal natural conditions, industrial cities can employ political and economic tactics to attract the many flourishing renewable energy companies and form a modern, green economy.

Hales, Roy L. "Hamburg is the gateway for Germany's Offshore Wind Industry." Oct 9, 2014. [http://cleantechnica.com/2014/10/09/the-gateway-for-germanys-offshore-wind-industry/].

Maunder, Hilke. "10,000 new jobs expected to be created by renewable energies." March 24, 2014. [http://www.hamburg-news.hamburg/en/cluster/renewable-energy/10000-new-jobs-expected-be-created-renewable-energ/].

Montgomery, James. "Germany's Wind Energy Nexus: A Tour Around Hamburg." June 2, 2014.
[http://www.renewableenergyworld.com/rea/news/article/2014/06/germanys-wind-energy-nexus-a-tour-around-hamburg].

Renewable Energy Hamburg. "Renewable Energy Hamburg: A Dedicated Industry Network." Oct 2012. [RenewableEnergyHamburg].

Obama Makes Moves to Advance Clean Energy in Place of Dirty Fuel
by Liza Farr

During his State of the Union speech this year, President Obama stated that he wanted Americans to win the race in innovation that spurs new jobs in industries of the future (Deese, Feb 10, 2015). In the past couple weeks, Obama has been making small but notable actions toward realizing this goal, through a new initiative supporting investment in clean energy, and discouraging dirty energy through the veto of the Keystone XL Pipeline. At the Advanced Research Projects Agency- Energy (ARPA-E) Summit, held from February 9 to 11, the White House announced the new Clean Energy Investment Initiative (Office of the Press Secretary, Feb 10, 2015). The goal of the project is to "catalyze $2 billion of expanded private sector investment in solutions in climate change, including innovative technologies with breakthrough potential to reduce carbon pollution" (Office of the Press Secretary, Feb 10). The initiative would help clean energy investors, including foundations, university endowments, and institutional investors, by reducing transaction costs, spreading promising investment models, and increasing their climate mitigation impact (Office of the Press Secretary, Feb 10, 2015). The White House will also be hosting a Clean Energy Investment Summit this spring to provide a forum for investors to collaborate and share best practices (Office of the Press Secretary, Feb 10, 2015). A number of philanthropic and private sector institutions have already made announcements that they are scaling up investment in clean energy innovations and technologies, including the University of California Board of Regents, the William and Flora Hewlett Foundation, the Schmidt Family Foundation, and Wells Fargo (Office of the Press Secretary, Feb 10, 2015).

On February 24, 2015, President Obama officially vetoed the bill authorizing the construction of the Keystone XL pipeline (Eilperin and Zezima, Feb 24, 2015). He reasoned that the bill cut short the consideration of issues of security, safety, and environment, and the effect of the pipeline on the national interest (Eilperin and Zezima, Feb 24, 2015). This is only the third veto Obama has ever issued, making it a significant statement (Eilperin and Zezima, Feb 24, 2015). In an interview with WDAY's Kerstin Kealy, Obama also noted he was happy to examine ways to increase pipeline production for US oil, but that the Keystone pipeline would be for Canadian oil (Glass-Moore, Feb 26, 2015). He said the pipeline would only create a little over 300 permanent jobs, and instead he wants to focus on infrastructure for American jobs and producers (Glass-Moore, Feb 26, 2015). Proponents of the President's move put a more environmental spin on it: many said approving or rejecting the pipeline is akin to a vote on climate change, and that the veto tells the world that the US is taking global warming seriously (Eilperin and Zezima, Feb 24, 2015). Although the President's messages on both the Keystone pipeline veto, and the Clean Energy Initiative focus more on the opportunity for advancing American jobs, investments, and technologies, the actions themselves speak louder. These are two significant actions for the US president to take toward mitigating climate change through advancing the clean energy industry in place of the fossil fuel industry.

Deese, Brian. "Increasing Investment in Clean Energy Technologies." February 10, 2015. [http://www.whitehouse.gov/blog/2015/02/10/increasing-investment-clean-energy-technologies].

Eilperin, Juliet and Zezime, Katie. "Obama Vetoes Keystone XL Bill." February 24, 2015. [http://www.washingtonpost.com/blogs/post-politics/wp/2015/02/24/keystone-xl-bill-a-k-a-veto-bait-heads-to-presidents-desk/?hpid=z1].

Glass-Moore, Adrian. "Obama explains Keystone XL pipeline veto in interview with WDAY's Kealy." February 26, 2015. [http://www.inforum.com/news/3688349-obama-explains-keystone-xl-pipeline-veto-interview-wdays-kealy].

White House Office of the Press Secretary. "FACT SHEET: Obama Administration Announces Initiative to Scale Up Investment in Clean Energy Innovation." February 10, 2015. [http://www.whitehouse.gov/the-press-office/2015/02/10/fact-sheet-obama-administration-announces-initiative-scale-investment-cl].

Texas Town Move to 100% Renewable Energy Bodes Well for American GHG Reduction
by Liza Farr

Georgetown is a seemingly average community of 50,000 people in the notoriously red state of Texas 25 miles north of Austin (Dart, March 2015). Over 40% of the town is over the age of 50, and it generally leans conservative. It's known for its old town charm and beautiful public square (Dart, March 2015). However, come January 2017, Georgetown will be the first state in Texas to be powered completely by renewable energy (Dart, March 2015). Jim Briggs, the interim city manager, was a key decision maker in the process, and wanted to clarify his motives: "we didn't do this to save the world- we did this to get a competitive rate and reduce the risk for our consumers" (Dart, March 2015). The city utility company in Georgetown has a monopoly on energy, which allows them to make the longer-term investments that are required for renewable energy facilities (Dart, March 2015). In Texas, as well as in many places, renewable energy is now cheaper than non-renewable energy (Dart, March 2015). After an analysis by city officials revealed renewable energy was cheaper for Georgetown in particular, they signed contracts with SunEdison and EDF Renewable Energy (Dart, March 2015). The agreement with EDF will give the town 144 MW of wind power from a future project near Amarillo that will be producing wind energy in January 2016 (Dart, March 2015). SunEdison will build solar plants in western Texas to provide the city with 150 MW of energy, completing the 100% renewable energy portfolio (Dart, March 2015). On average, one megawatt of energy can power 100 Texas homes on a hot summer day, and even more in other seasons (Malewitz, March 2015). Since the wind is strongest after the sun has gone down, the two types of energy complement each other well, and the town predicts they will only need to use the backup conventional power during peak demand and during some of the summer (Ross, March 2015). The city will end up feeding into the grid more energy than they use, so other power users in the state will reap the benefit as well (Ross, March 2015).

The daring move by city officials has made residents happy primarily because of it will give them the security of a fixed rate plan similar to the current cost of 9.6 cents for kilowatt hour (Dart, March 2015). Additionally, using renewable energy protects against fluctuations in oil and gas prices, which are predicted to increase (Dart, March 2015). Businesses have already been talking with the local government about making moves to Georgetown, both for the cheaper energy, and to boost their green credentials, an increasingly important aspect of PR (Dart, March 2015).

There are many reasons why the state is a good candidate to move toward renewable energy. The western region of Texas near New Mexico is one of the most promising solar resource site in the nation, and Texas is also the nation's largest wind power producer (Dart, March 2015). Renewables also use much less water, which is needed in the drought stricken state (Dart, March 2015). Although some factions of the state government have denied a link between greenhouse gas emissions and climate, Texas has spent 7 billion dollars on the Competitive Renewable Energy Zone, which built 3,600 miles of transmission lines to connect western Texas wind power to major urban areas (Ross, March 2015). Overall, Texas has failed to provide incentives for renewable energy, which has left the market lagging behind many states with less renewable energy resources (Malewitz, March 2015). The state also released a report in September that concluded renewable energy was not reliable or extensive enough to meet peak demand (dart, March 2015). Despite some resistance in the state government, Texas is clearly moving toward embracing renewable energy. Last May, Austin Energy signed an agreement with a California contractor to build a 150 MW solar farm (malewitz, March 2015). Now, Georgetown joins the few US cities, including Burlington, VT, Palo Alto CA, and Aspen, CO, in going 100% renewable (Malewitz, March 2015). If a state notorious for being fossil fuel loving is beginning to shift to renewable energy, prospects look good for the spread of renewable energy in America.

Malewitz, Jim. Georgetown goes all Renewable Energy. March 18, 2015.
[http://www.texastribune.org/2015/03/18/georgetown-goes-all-renewable-energy/]
Dart, Tom. Texas city opts for 100% renewable energy – to save cash, not the planet. March 29, 2015.
[http://www.theguardian.com/environment/2015/mar/28/georgetown-texas-renewable-green-energy]
Ross, Dale. Mayor: Why My Texas Town Ditched Fossil Fuel. March 27, 2015.
[http://time.com/3761952/georgetown-texas-fossil-fuel-renewable-energy/]

Energy Star Initiative
by Dylan Goodman

Energy Star is a volunteer US Environmental Protection Agency (EPA) program dedicated to helping businesses and individuals save money and reduce greenhouse gas emissions. Originally introduced in 1992 under the Clean Air Act, the program works as a means of developing, promoting, and increasing the use of energy efficient products to reduce carbon emissions. The first products to gain Energy Star certification were computers and monitors before expanding to include office appliances and residential heating and cooling systems. By 1996 the Energy Star program had partnered with the US Department of Energy to certify more major home and industrial appliances. The Energy Star program has become increasingly successful, saving business and home owners over $24 billion in 2012 alone.

The most recent appliances to join the ranks of Energy Star certification are common household dryers. The Energy Star program has become increasingly inclusive, and the addition of Energy Star certified dryers marks the first time in 6 years a major household appliance has been added. Beginning this past presidents day weekend 45 new Energy Star certified dryers are now available on the market. While refrigerators were once some of the most power hungry appliances, they have been made much more efficient in recent years. Today, "Dryers are one of the most common household appliances and the biggest energy users." (Gina McCarthy, EPA). In order to further energy saving initiatives, it is imperative that these high power-consumption devices be made more efficient. Current Energy Star certified dryers include both electric and gas models across 5 household brands. Rather than relying on timed drying periods, all models use moisture sensors to ensure that they don't continue to run after clothes have been dried. This alone is estimated decrease energy consumption by 20% according to the EPA. Some of the more advanced models incorporate a heat pump system, a process that allows them to recycle heated air. This technology is estimated to increase efficiency by another 40%. Although Energy Star certified dryers cost roughly $600 more than conventional dryers, this cost is offset by government rebates and future savings. The EPA estimates that if all dryers sold in the US were Energy Star certified, energy savings could reach $1.5 billion while reducing greenhouse gas emissions equal to that of 2 million vehicles, annually.

"History." Energy Star. N.p. Web.

Roth, Katherine. "Energy-hogging Dryers Go Green with New Energy Star Cred." The Denver Post. N.p., 21 Feb. 2015. Web.

Starting Small: Shifting into Renewable Energy in South Korea

by Alison Kibe

In 2009, South Korea began its foray into smart grid technology through a $220 million pilot project to implement the technology into about 2,000 homes on the coast of the island of Jeju. The project served as a test bed for home electricity communication with power suppliers, the purpose of which was to learn how the grid could be implemented among larger populations (McDonald, 2011). Today, South Korea is moving more and more toward clean energy sources. Korean legislation dictates that at least 2% of gross domestic product must be spent on the research and development of renewable energy and by 2022, companies must have 10% of their power come from renewables (McDonald, 2011). To prepare for this deadline, Korea is once again using small populations to test out renewable systems, this time looking to small island populations.

The island of Gasa sits off the southern coast of mainland South Korea and has a population of 286 people. Diesel generators were used to supply energy to the island community, but after the installation of wind turbines, solar panels, and energy storage across the island in October of 2014, the 168 island households of Gasa now have access to their own renewably powered micro grid (Kim, 2015). Micro grids have become an attractive option for small and more isolated rural communities due to their size and ability to disconnected from larger grids if there are energy disruptions. This option is turning out to be cheaper than using diesel fuel that had to be transported to the island with expected energy cost reductions at nearly $300,000 per year (Kim, 2015).

The informational outcomes of the Gasa project will be used to scale it to work in larger populations and to implement micro grid projects to other isolated, rural, and developing regions. Korea has set the goal of expanding its micro grid projects into 120 other islands (Kim, 2015). Just doing this is expected to reduce costs by $14.5 million and reduce greenhouse gas emissions by 80% among the islands (Kim, 2015). With these levels of savings and improvements in efficiency, there will be an opening of economic opportunities in rural areas, which is often where they are most needed.

McDonald, Mark. "To Build a Better Grid." The New York Times. July 2011. http://www.nytimes.com/2011/07/29/business/global/to-build-a-better-grid.html?_r=1&ref=smartcities

Kim, Sungwoo. Clean energy + technology = resilient energy infrastructure. Devex. April 2015. http://news.uark.edu/articles/25992/picasolar-raises-1-2-million-in-equity-investments

A Greener Apple
by *Briton Lee*

In February 2015, Apple has announced at an investors' conference that it has moved to build a solar farm in Monterey County to power all of its operations in California, including headquarters, data centers, and stores. This agreement with the largest US developer of solar farms, First Solar, is the largest investment in solar power ever by a non-utility company. The projected amount of energy that can be obtained from this solar farm is 130 megawatts, which could be used to power 60,000 homes (Randall 2015). Since its rating as the worst tech offender by Greenpeace in 2011, Apple has made good on its decisions to improve its environmental footprint (Levy 2014). CEO Tim Cook hired the previous Administrator of the US EPA, Lisa P. Jackson, as its Vice President of Environmental Initiatives in 2013, a move integral to its improvement (Levy

2014). The title 'Environmental Initiatives' underlines the fact that Apple wants to be a company leading the efforts in being environmentally conscious. Cook notes that he wants to inspire other major tech companies to be more cognizant of their environmental impact (Chen 2015). In fact, other tech giants are recognizing the reality of climate change, with Google, Microsoft, and Amazon all having large investments in wind power. However, it is not so easy to paint these tech companies as altruistic; the "halo effect" of appearing environmentally conscious translates directly into profits, as evidenced by the fact that immediately after announcement of the investment, Apple became the first US company to end the trading day with assets over $700 billion. Additionally, the agreement should lead to completion of the project by the end of 2016, which is just in time to take advantage of a 30% tax credit (Randall 2015). This is not to say that such reasons for investment in renewable energy is bad, and incentivizing green energy by making it economically profitable may be a strong way to work toward a truly sustainable future.

Randall, Tom. 2015. Bloomberg Business. "What Apple Just Did in Solar Is a Really Big Deal". (http://www.bloomberg.com/news/articles/2015-02-11/what-apple-just-did-in-solar-is-a-really-big-deal)

Levy, Steven. 2014. Wired. "Apple Aims to Shrink Its Carbon Footprint With New Data Centers". (http://www.wired.com/2014/04/green-apple/)

Chen, Brian. 2015. New York Times. "Apple Building Solar Farm to Power California Operations". (http://bits.blogs.nytimes.com/2015/02/10/apple-to-build-california-solar-farm/?_r=1)

Obama's Solar Energy Initiative
by Shannon O'Neill

On April 3rd, 2015, President Obama visited Hill Air Force Base in Utah to announce a Solar Energy Initiative. Since Obama took office in 2008, solar energy production in the US has increased 20-fold. Additionally, as the cost of solar manufacturing and research has decreased by more than 12%, demand for solar energy has increased, allowing this industry to provide more jobs to the economy than other sector. These facts led to Obama's announcement to address climate change by creating more jobs, specifically for veterans, in the clean energy sector.

This initiative stated that the Department of Energy (DOE) set a goal to train 75,000 solar energy workers by 2020. Additionally, the DOE partnered with the Department of Defense to create a "Solar Vets" program at 10 military bases across the nation, including Hill Air Force Base. This project focuses on training veterans to be solar energy workers. The training is based specifically on technician and work-based strategies that such personnel acquired through their time in the military. The veterans will learn how to install solar panels, connect electricity to the gird, and understand the local building codes in an intensive 4 to 6 week session.

The pilot "Solar Vets" program proved to be a success, as all 20 veterans who went through the training received a job offer from one of the top 5 solar firms. This program proves to be vital in easing the transition to civilian life for veterans while further increasing interests and fulfilling demands for solar workers in the solar energy sector.

In addition to this program, the military has been involved with other solar energy initiatives. Hill Air Force Base has already installed solar panels, and this week the Army began a 15-megawatt solar project at Fort Detrick in Maryland, which has the capability to provide energy for 2,500 homes. Obama's initiative will set the nation up to "cut net greenhouse gas emissions by 26-28% below 2005 levels by 2025" (Fox News).

Fox News: "White House releases clean energy fact sheet ahead of Pres. Obama's speech in Utah" (http://fox13now.com/2015/04/03/white-house-releases-clean-energy-fact-sheet-ahead-of-president-obamas-speech-in-utah/)
Washington Post: "Obama Touts Ambitious Solar Jobs Initiative in Conservative Utah (http://www.washingtonpost.com/politics/obama-touts-ambitious-solar-jobs-initiative-in-conservative-utah/2015/04/03/70ec8dce-da12-11e4-8103-fa84725dbf9d_story.html_)

Obama and Modi Negotiate Renewable Energy in India
by Melanie Paty

On January 25th, 2015, President Obama met with newly elected prime minister of India, Narendra Modi, primarily to discuss climate change. Ari Philips published an article for Think Progress that gives context for the negotiations and explains the climate conditions India is currently facing. The article cites staggering statistics about the urgency of pollution mitigation in India. Delhi is the most polluted city in the world with PM2.5 levels eight times higher than EPA approved levels and air pollution-related ailments contributing to 109,000 premature adult deaths per year. Delhi is not the only problem as India houses thirteen of the twenty most polluted cities. A chance to turn to a more sustainable path has opened with Modi taking office as he has increased India's 2020 solar energy target by five times. The article sites Raymond Vickery who notes that this ambitious goal will be financially difficult for India as its new climate goals are estimated to require $100 billion in investments, much of which will need to come from private investors.

In a recent Huffington Post Article, Manish Bapna agrees with Vickery saying that securing an investment flow to support India's energy targets is one of four suggested steps he recommends for building a successful climate change strategy between the United States and India. The other three steps he recommended were combining to lead development of global clean energy collaborative, increasing efforts climate change education initiatives in both countries, and establishing a direct line of communication between the two governments to best prepare for a new international climate agreement in Paris 2015. The probability of these steps coming to fruition is unclear. However, private investment flow from the United States is already looking promising as US-based SunEdison is developing renewable energy in Karnataka and a $4 billion solar manufacturing facility.

Philips, Ari. "In the World's Most Polluted City, Obama Seeks Climate Ally." Think Progress. January 22 2015.
(http://thinkprogress.org/climate/2015/01/22/3614061/obama-goes-to-india/)
Bapna, Manish. "As Obama and Modi Meet, 4 Opportunities for US-India Action on Clean Energy and Climate Change. Huffington Post. January 22 2015.
(http://www.huffingtonpost.com/manish-bapna/as-obama-and-modi-meet-4-_b_6519026.html)

California's Investment in Clean-Tech is Paying Off

by Melanie Paty

A recent article on Renewable Energy World by Bloomberg's Mark Chediak and Miachel B. Marois proves that Governor Jerry Brown's goal to "show that decarbonizing is consistent with economic abundance financial stability" is becoming a reality. Since Brown came to office in 1975, he has been pushing for policies that support green technology and today, California has the most ambitious renewable energy goals in the nation. Within the next 15 years, Brown has proposed increasing electricity supply by renewables from 30% to 50%, reducing petroleum use in cars and trucks by 50%, increasing energy efficiency of existing buildings by 200%, and putting 1.5 million zero emission cars on the roads. While most governments are hopping on the sustainability bandwagon in recent years, California's early investment seems to be paying off.

The article examines the performance of the 26 California-based clean-tech companies relative to other stocks within the NYSE Bloomberg Americas Clean Energy Index. The Californian companies are expected to climb 49% higher than the others in the next year. They also spend an average of $141 million on research and development, which is approximately 25% of sales, whereas other US companies in the index spend an average of $85 million, only 10% of sales. Furthermore, the Californian companies employ a total of 431,000 people across the nation; Tesla and SolarCity, two of the biggest California-based firms in the index, have been generating jobs at a rate of 9.5% per year over the past two years. Additionally, over the past eight years, investors have provided more than $27 billion dollars in financing for clean technology in California through venture capital and other avenues. Opposition argues that expansion of renewable energy will push energy prices up when California already faces 79% higher rates than the rest of the Unites States due to the inclusion of renewables and gas.

Renewable Energy World
(http://www.renewableenergyworld.com/rea/news/article/2015/02/californias-clean-tech-industry-best-in-us-for-jobs-and-investment)
TAGS: Jerry Brown, California, Clean Technology, Renewable Energy, Bloomberg
Tweet: Jerry Brown is the man for going green in the 70s! Bloomberg writers report that California's clean technology companies are the best in the US for job creation and Investment! @JerryBrown @RenewableEnergyWorld @Bloomberg

Effects of Utility Scale Solar Energy on Aquatic Ecosystems in the Southwest
by Melanie Paty

In a recent Environmental Management article, Grippo, Hayse, and O'Connor (2015) speculate on the potential detriments solar farm development imposes on temporary bodies of water and the wildlife that depend on them. The authors' locational focus is the Southwestern United States, where there are more than 40 pending or approved solar development permits. Temporary bodies of water, either intermittent, "seasonally dry stream, especially during times of low rainfall or high heat," or ephemeral, "defined as those that do not receive groundwater inputs and contain water only briefly and in direct response to precipitation," play an important role in desert ecosystems: they connect the landscape, transport water and nutrients downstream, serve as the short-term habitat for animals with aquatic life stages and the reproductive site for various animals, and give rise to riparian corridors that provide a variety of benefits to local animal species. However, unlike permanent bodies of water, temporary water is not protected under the Clean Water Act, unless it is significantly connected to a permanent body of water. Thus, temporary bodies of water are often disturbed in construction or operation of solar farms; the three greatest concerns are disrupted flow, water contamination, and groundwater depletion. Water quality is jeopardized because in order to limit the amount of dust that accumulates on the mirrors and panels, solar farms use dust suppressants that contain chloride salts and brines. Limited data are available on the toxicity of these substances, but some have been shown to increase algal growth, reduce oxygen levels, and increase salt content to a point that is toxic for aquatic species. The authors recognize that it is impossible to go forward with solar plant construction without disturbing the temporary water bodies, so they suggest minimizing the impacts in a variety of ways, primarily through increased monitoring of detrimental effects. Some examples include: finding the sustainable groundwater yield, calculating the vulnerability of streams to erosion, and mapping groundwater-dependent communities at regional and site-specific levels.

While increase monitoring would be helpful, it would also be interesting to do a cost benefit analysis of the monitoring and avoidance measures, because, as the authors note, none of the threats have proven to produce population-level effects. While solar farms do have negative environmental impacts on the desert ecosystem, which is undervalued, the oil and gas industry has also had detrimental effects on many other ecosystems; thus, while it is important to ensure that solar farms are employing the most sustainable practices, it is also important to recognize that there are always tradeoffs, and clean energy offers much promise for a more sustainable future overall, despite some negative effects.

Grippo, M., Hayse, J., O'Connor, B. (2015). Solar Energy Development and Aquatic Ecosystems in the Southwestern United States: Potential Impacts, Mitigation, and Research Needs. Environmental Management (55) 244–256.

Clean Energy in India's 2015–2016 Budget Proposal
by Melanie Paty

On February 28 2015, Finance Minister Arun Jaitley announced India's fiscal budget proposal and critics were disappointed by the lack of definitive renewable energy plans.

Jaitley reiterated the 2022 clean energy target of 1,750,000 megawatts consisting of 1,000,000mW solar, 60,000 mW wind power, 10,000mW biomass, and 5,000mW hydroelectric. As of December 2014, India has only approximately 34,000mW of renewable energy capacity and it is estimated that $40.5 billion dollars of infrastructure investment is needed to reach its targets. However, the budget does not incentivize investment or manufacturing in the sector. One of the few allocations Jaitley announced is the opening of a national infrastructure investment fund for which they intend to secure $3.2 billion of inflow per year; however, this fund will finance many infrastructure projects, not only those pertaining to renewable energy. The only direct financing for clean energy is an excise duty cut on copper wire and tin alloys, which are used to make PV cells, but according to a consultant the impact on cost of production will be negligible. Jaitley also announced an allocation of $122 million for a 2015–2016 program designed to speed up the manufacturing process for electric cars. Anish De, KPMG India Infrastructure and Government Services Partner, notes that no tax free bonds have been specifically proposed for renewable energy, so any funding from bonds will have to come out of generalized infrastructure bonds. Jaitley also announced an increase in the coal cess in order to help sustain subsidies for clean energy projects, but no specific plans for the utilization of such funds was delineated. It is surprising how little of the budget was dedicated to renewable energy incentives given Modi's stance on solar energy. The lack of government investment in clean energy suggests that it is relying on private investment to fuel the sector, which has set a target of $200 billion for foreign and domestic investments. India should dedicate more government funding to the clean energy sector as it is projected to generate $160 billion of business opportunities over the next five years.

Upadhyay, Anindya. "Renewable energy sector upbeat on Budget 2015 proposals." The Economic Times. February 28, 2015.
(http://economictimes.indiatimes.com/industry/energy/power/renewable-energy-sector-upbeat-on-budget-2015-proposals/articleshow/46409521.cms)

G. Balachandar. "Budget fails to cheer clean energy sector." The Hindu. February 28, 2015. (http://www.thehindu.com/business/budget/budget-fails-to-cheer-clean-energy-sector/article6946291.ece).

Greentech Lead. "Budget 2015: No incentives but bigger targets for renewable energy." February 28, 2015. (http://www.greentechlead.com/news/budget-2015-no-incentives-but-bigger-targets-for-renewable-energy-22027).

Bhat, Shravan. "Budget 2015 lukewarm to the green sector." Forbes India. February 28, 2015. (http://forbesindia.com/article/budget-2015/budget-2015-lukewarm-to-the-green-energy-sector/39741/1

Costa Rican Electricity in Q1 2015 has been 100% Renewable

by Melanie Paty

Between the first of the year and March 2015, Costa Rica has functioned on electricity by 100% renewable. Costa Rica boasts an impressive electricity portfolio, one source reporting 80% hydro powered and 13% geothermal and another reporting 88% total renewable energy with 68% hydro, 15% geothermal, 5% wind, and a small percentage of solar and biomass. Heavy rains at four hydroelectricity plants throughout the first quarter of 2015 allowed for the country to meet this impressive renewable energy milestone. However, the government is pushing geothermal development to ensure the renewable energy supply is resilient to climate change as water supply could be reduced in future years. Costa Rica is well endowed with geothermal energy with many volcanoes, and in mid-2014 the government invested $958 million in a geothermal project that is expected to produce approximately 210 MW. Gough (2015) notes that the move towards geothermal is also environmentally beneficial as hydropower has several questionable environmental downsides including changes to fish migration patterns. It is important to note that this milestone, while impressive, has been easier to maintain in Costa Rica than other countries for two main reasons. First, it is a very small country about half the physical size of Kentucky with a population of only 4.8 million. Second, it has been able to support costly environmental policies because of excess government funding available after the disbandment of military in 1948. Costa Rica is known for its strong environmental policy platform including a goal to be carbon neutral by 2021, the use of fees and taxes to support land preservation, and a reward system in which landowners are paid to plant trees and to not cut down old growth forests.

Gough, Miles. Science Alert, "Costa Rica powered with 100% renewable energy for 75 straight days" March 20 2015. (http://www.sciencealert.com/costa-rica-powered-with-100-renewable-energy-for-75-days).

Inquisitr, "Running Entire Nation Solely On Renewable Energy: Costa Rica Powered With 100 Percent Eco-Friendly Sources For 75 Straight Days" March 20 2015. (http://www.inquisitr.com/1941378/running-entire-nation-solely-on-renewable-energy-costa-rica-powered-with-100-percent-eco-friendly-sources-for-75-straight-days/)

Valentine, Katie. Think Progress, "Costa Rica Has Gotten All Of Its Energy From Renewables For 75 Days Straight" March 21 2015. (http://thinkprogress.org/climate/2015/03/21/3636823/costa-rica-renewables/)

Delays in the Desert Renewable Energy Conservation Plan

by Melanie Paty

Federal and state officials have spent over six years discussing land allocation for the Desert Renewable Energy Conservation Plan in the Mojave Desert. To simultaneously satisfy both conservation and renewable energy development agendas, the government has tried to target the least environmentally sensitive land for development and the original plan was to place 80% of development on 2 million acres of private land. Now, 22 million acres have been incorporated into the plan: approximately 50% public and 50% private. Recently, due to criticism from several local counties, projects on the private land have been delayed. However, the Bureau of Land Management (BLM) plans to move forward with development on the public land.

The counties dissatisfied with the plan are Los Angeles, San Diego, Imperial, Kern, Riverside, San Bernardino and Inyo. These counties have permitting authority on the private land in question and have voiced several criticisms including concerns that the plan conflicts with existing environmental programs, fear of tax revenue loss because California has tax-exempted solar projects, and they want a more detailed cost benefit analysis performed before development commences. The government has decided to resolve these issues on a county-by-county basis, hence the delay in development. However, some fear this decision will encourage the BLM to open up more public land for development and public land is far more pristine and worth conserving than the private land in question.

Casto and Dashiell of the Victorville Daily Press note that most local communities approve of the decision to delay development on private lands, but hope the BLM stays true to its original renewable energy development and conservation goals and does not make more public land available for development. The LA times editorial board states that the BLM should hold firm as well, suggesting a preferable strategy would be speeding up the private land allocation process by offering incentives to make up for tax revenue loss in the counties.

Sahagun, Louis. LA Times, "Desert Renewable Energy Plan is Altered to Win Counties' Support." March 11 2015 (http://www.latimes.com/local/california/la-me-0311-desert-20150311-story.html).

LA Times, "An Unnecessary Delay for Mojave Desert Renewable Energy Plan." April 1 2015 (http://www.latimes.com/opinion/editorials/la-ed-desert-renewable-energy-conservation-plan-mojave-20150401-story.html).

Castro and Dashiell. Victorville Daily Press, "BLM Shouldn't Change its Goals for Renewable Energy or Conservation." March 30 2015

(http://www.vvdailypress.com/article/20150330/NEWS/150339966/12967/LIF
ESTYLE).

The Future of Climate Change in Hilary Clinton's 2016 Campaign

by Melanie Paty

A recent Guardian article by Gambino and McCarthy speculates on how Hillary Clinton's 2016 presidential campaign will address climate change. Environmental groups are anxious to see what is in store after John Podesta, Clinton's campaign chairman, tweeted "Helping working families succeed, building small businesses, tackling climate change & clean energy. Top of the agenda. #Hillary2016." Podesta is a senior advisor to President Obama so his appointment as campaign chairman sent a strong signal to begin with. The article analyzes Hillary's past actions and public stances on environmental policy, revealing a disjointed track record. Evidence that suggests she might support a strong climate change mitigation strategy include her remarks at the National Clean Energy Summit last September in which she described global warming as the most urgent issue of our time. Also, in December, she mentioned the importance of protecting Obama's climate change policies and The Clinton Foundation recently stopped accepting donations from several Gulf States. However, other pieces of evidence point to a less promising stance. The article quotes Bill McKibben stating that "she was the world's top diplomat when the climate talks in Copenhagen fell apart. And she's awfully cozy with big oil and gas interests." Additionally, in her last presidential campaign she supported a gas tax holiday and led an initiative to promote fracking across the world. Skeptics like McKibben are hoping that Clinton comes out with concrete strategies like imposing new carbon caps or opposing the Keystone pipeline. Regardless of the specificity, recent polls show that supporting federal action against climate change could be extremely beneficial to the campaign. A Washington Post and ABC news poll found that a majority of voters are "very" or "extremely" supportive of a president who intends to take federal action against climate change, and a similar poll conducted by Resources for the Future, Stanford, and The New York Times found that the majority of voters, including republicans, think the government should play a role in climate change mitigation. Thus far, according to the evidence in the article, Clinton's potential position on the issue is unclear. The authors note that since her campaign launch, Clinton has yet to say the words 'climate change.' No concrete evidence will arise until the summer when her full climate change agenda is released. For now environmental groups will have to keep a close eye on the media in the upcoming months to get a clue as to where Hilary might be headed.

Lauren Gambino and Tom McCarthy. "Hillary Clinton's green path to the White House: will she be 'careful' on climate?" The Guardian April 20, 2015 (http://www.theguardian.com/us-news/2015/apr/20/hillary-clinton-white-house-climate-change)

Hawaii Set on Being the First State to Run on 100% Renewable Energy
by Melanie Paty

On April 24, 2015 in The Hill, Joshua Reichert provided a very positive outlook on the future of the clean energy sector and called on the government for significant investment. He notes the tremendous growth the sector has recently experienced following federal incentive programs such as the tax credit for solar and other support programs for wind: solar capacity has more than doubled since 2008, the average price of panels has declined 63%, wind farms generate enough energy to supply 15 million homes and 500 domestic manufacturing facilities, and the two economies combined employ over 200,000 people. Reichert urges the government to continue to support the growth of the clean energy sector in this fashion and is hopeful that both parties are willing to collaborate on this issue given the recent passage of the Energy Efficiency Improvement Act of 2015. He acknowledges that at this point, a lot of the exciting innovation lies with small start-ups that need government support to grow, but he urges us to remember that at one point the oil and gas industry was composed of similarly sized ventures and grew to be one of the most profitable sectors after government investment. In his opinion, there is virtually no downside to investment in the clean energy sector as it reduces emissions, creates jobs, increases the tax base, and can help give the United States a competitive advantage, which, as of now, we are lacking. In order to stimulate the kind of investment he hopes to see, he urges the government to approve the $2.7 billion budget allocation for the Department of Energy Obama requested for 2016 and to continue to incentivize private investment in renewable energy.

The urgency with which Reichart speaks is warranted given the historical fluctuations of government investment in energy research and development. In order to become leaders in the clean energy race, the United States government will have to pass legislation that includes programs like a feed-in tariff and increased government grants for demonstration plants, which have been successful in stimulating the sector in Europe. While private sector investment is critical in mitigation climate change in a timely fashion, in this early stage of the sector, clean energy investment is highly risky and could benefit from increased federal investment to take some of the burden off of private investors.

Reichert, Joshua. "Renewable Energy, the Best Investment of the 21st Century." The Hill, April 24th 2015 (http://thehill.com/blogs/pundits-blog/energy-environment/239921-renewable-energy-the-best-investment-of-the-21st)

How far have we come?
by Ali Siddiqui

An article by Chris Mooney for The Washington Post attempts to determine whether the United States of America really is changing the way it uses energy. The conclusion of the article was that America has changed. Mooney gains most of his evidence for America's change from the Sustainable Energy in America Factbook prepared by the Business Council for Sustainable Energy and published by the Bloomberg New Energy Finance.

Chris Mooney doesn't take America too far back in the past to analyze whether a change has occurred. He compares a 2014 America to a 2007 America. By citing the Factbook, Mooney shows how, although America fell short to China in having the largest investment into clean energy, it was still able to manage nearly $400 billion in clean energy investment. He, also, shows how even after the economic bounce-back from the recession in 2008-2009, which some might have forecasted to mean greater American energy expenditure, America actually maintained a flat electrical demand alongside a decline in overall carbon emissions (9%).

Even more incredible is America's skyrocketing growth in two new energy sectors; since 2008 wind and solar energy have seen a growth of more than three times in capacity. In 2007, Bloomberg New Energy Finance and the Business Council for Sustainable Energy reported renewable energy to have provided 7% of America's total energy. With wind and solar energy becoming a lower cost option, many homeowners may witness a change in how their neighborhoods are receiving energy. Needless to say, America is moving forward on its green initiative. One simply needs to look at the past, a 2007 America, to see how far we've come.

Mooney, Chris. Report: Wind and Solar Energy have tripled since 2008. Washington Post. February 4th 2015. http://www.washingtonpost.com/news/energy-environment/wp/2015/02/04/report-wind-and-solar-energy-have-tripled-since-2008/

Apple and Google put their Green Thumbs Down
by Ali Siddiqui

Facebook, Amazon, Salesforce, and Yahoo are a part of a the latest wave of technological companies that have increased their spending on renewable energy by doubling down on their investments. Matt O' Brien and Julia Love have written an article in The San Jose Mercury News, which shows that this increased spending is nothing compared to that of Apple and Google.

Apple and Google have raised the stakes in investing in renewable energy by not just dipping their toes into their budget, but by jumping into the water headfirst. Recently, Apple has agreed to an $848 million deal that utilizes a remote Monterey county valley along the San Andreas Fault in order to lay down solar panels in a 280-mega watt solar farm. Apple's C.E.O Tim Cook revelation of this deal at a San Francisco tech conference not only pleased shareholders, but also raised Apples' market value to $700 billion. Google, which has already spent $1.5 billion on renewable energy across the world, is also making a move towards investing in renewable energy closer to its corporate bases. Google's new 20-year deal with Florida-based Company NextEra looks to build an 86-mega watt farm by the end of this year in Alameda. Apple is even stating that these investments in renewable energy closer to home will generate enough energy to power all 52 Californian stores.

Critics of these investments are skeptical of these companies' true intentions. Adam Browning, an executive director of clean energy advocate Vote Solar, has pointed out that by heavy green investment, these companies are taking measures to protect themselves from the finicky changes in price of oil, natural gas, and other natural resources. Another controversial issue surrounding these deals is that Apple's new investment will replace 1,300 acres of grassland currently populated by wildlife. Google's investment, however, re-utilizes an existing wind warm. Google plans on removing wind turbines that number around 770 and were built in the 1980s. They plan to replace them with 48 wind turbines of new design that will double their energy output. Regardless of the intentions of these companies, technology companies are making huge moves in the renewable energy sphere. If this continues then it is possible that Silicon Valley will become not only a hub for technology, but also for green energy.

O'Brien, Matt., Love, Julia. Google and Apple harness wind and sun in huge green energy deals in to power Silicon Valley campuses. February 11th 2015. http://www.mercurynews.com/business/ci_27504196/google-and-apple-harness-wind-and-sun-huge?source=infinite

Costa Rica Using Renewable Energy only for Past 75 Days

by Ali Siddiqui

Costa Rica is a small country with a population of approximately 4.8 million people and since December 2014 electricity has been generated without a single use of a fossil fuel. Past heavy rainfalls have powered four hydroelectric plants within the first three months of the year, which has been sufficient to power the country. This accomplishment has pushed Costa Rica one step closer to its goal of becoming carbon-neutral by 2021.

Although, this incredible feat has been accomplished through the use of hydroelectric plants, 10% of Costa Rica's energy is generated by geothermal energy. With an eye forward, Costa Rica has begun investment into increasing that percentage and diverting from hydropower, which accounted for 80% of its energy last year. The reasoning behind Costa Rica's aim to diversify its alternative energy sources from just hydropower is the fear that being too dependent on hydropower leaves the country susceptible to large-scale consequences caused by drought and seasonal changes.

Currently 94% of Costa Rica's energy is made by renewable energy and the commitment to improve that percentage is strong. Costa Rica already does not exploit its rich oil deposits along the Caribbean coast. Another example of the Costa Rican government's commitment is the $958 million US dollars invested into geothermal energy back in mid-2014. The investment is supposed to allow for one geothermal plant to produce around 55 megawatts of electricity and two plants that can further provide 50 megawatts of electricity each. 55 megawatts of electricity can power approximately close to 55,000 homes.

The infrastructure of this Central American country is also highly conducive to alternative energy production and the World Economic Forum has recognized this fact and ranked Costa Rica second in Latin America for its electricity and telecommunications infrastructure.

Tufft, Ben. Costa Rica goes 75 days powering itself using only renewable energy. March 22nd 2015. http://www.independent.co.uk/news/world/americas/costa-rica-goes-75-days-powering-itself-using-only-renewable-energy-10126127.html

The Battle between Ohio and Solar Energy
by Ali Siddiqui

Governor John Kasich's recently-passed legislation Senate Bill 310 pauses Ohio's benchmarks for renewable energy and energy efficiency for two years. This legislation has been met with some controversy by solar power developers. Supporters of the bill expect this legislation will result in lower electricity bills for residents and businesses, however, dissenters believe that this pushes the development of solar energy out of state.

Hull and Associates of Dublin vice president Steve Giles and vice president of solar business development for Melink Corp. Jeremy Chapman are both moving outside of the Ohio area to look for projects to undertake. Chapman's company completed 90% of their solar projects outside of Ohio last year, and the strain of traveling has been hard on him and his family.

In 2012, Ohio saw the peak of solar energy installations. The next year the number of installations declined, however, the new projects were generating around 10.4 megawatts of energy less than those installed in 2012. Ohio's primary surge in solar power development came as a result of legislation passed in 2008, which had Ohio adopt renewable-energy benchmarks up until 2025. This had companies initiate huge projects to meet those requirements, which essentially are frozen now. The effects of the freeze can be seen by the lower value of solar energy credits currently. In 2011 solar energy credits peaked at the price point of $400, but now are priced at $48.

Ohio-based solar energy developers are suffering because they are now competing with local solar energy developers in out of state areas. However, Ohio Senator Bill Seitz stands by Ohio's decision on the new legislation. His argument is that a pause on the benchmarks and a decrease in electricity bills will cause a net increase in jobs. Solar power developers disagree.

Gearino, Dan. Solar Energy Gets Cloudy in Ohio. April 19th 2015.
http://www.dispatch.com/content/stories/business/2015/04/19/01-solar-cloudy.html

Quadrennial Energy Review
by Ali Siddiqui

The Obama administration recently released a report titled the "Quadrennial Energy Review." This report states that our methods of obtaining electricity needs "significant change", so that sectors like renewable energy and distributed energy (solar panels on rooftops) continue to grow.

The administration is currently attempting to limit greenhouse gas emissions by power plants. Solar and wind energy is rising. Energy from coal is not only declining, but also being overtaken quickly by energy from natural gas. These changes compounded with the report's notes about the dated energy infrastructure the US still possesses is pushing the Obama administration towards modernizing the US's energy infrastructure.

The Quadrennial Energy Review was started in 2013 after the Obama administration had passed the Clean Power Plan. The name of the review mimics the same review of the US's defense system's capabilities, "Pentagon's Quadrennial Defense Review". The approach of this review is different. It analyzes challenges that currently exist within the US's energy infrastructure as well as addresses the challenges posed by climate change. For example, the report indicates that one of the biggest vulnerabilities of the grid is the risk associated with loss of transformers and a way to circumvent that risk is having states analyze these risks and begin plans to mitigate them.

While there are those who voice opposition towards the Clean Power Plan and the resulting Quadrennial Review, Vice President Joe Biden, White House Science Advisor John Holdren, and US Energy Secretary Ernest Moiz support this report and will hold discussions about this report soon. Although, there is controversy about the direction the US should move in regarding energy, the findings of the report that are fact are the dated pipelines and infrastructure put in place for gas transmission and gathering. Close to half of the US's pipelines were built in the 1950's and 1960's. Regardless of whether or not we move towards modernizing our infrastructure for green energy or not, shouldn't this dated infrastructure still be re-evaluated?

Mooney, Chris. The US's energy infrastructure will need major changes, says Obama report. April 21st 2015. http://www.washingtonpost.com/news/energy-environment/wp/2015/04/21/major-changes-needed-for-u-s-power-infrastructure-says-obama-report/

Drought-Ridden California Looks to Renewable Energy to Keep the Water Flowing

by Trevor Smith

California's devastating drought means not only that there is less water available in the state overall, but also that what little water remains is increasingly difficult to retrieve, requiring much more energy in order to pump the water to where it needs to go. The problem is compounded by the fact that the hydropower some water utility companies rely on tends to dwindle in the face of severe drought (UCS 2015). However, some view this crisis as an opportunity for renewable, clean energy, hoping to meet the increased energy needs of water utilities through a mix of solar, wind, and geothermal energy. Sonoma County, just north of San Francisco, has already managed to power its entire water supply system with energy from renewable sources (Hart 2015).

One recent estimate released by the Union of Concerned Scientists (UCS) suggests that a full 20% of California's energy consumption is due to water pumping, treatment, and heating (Hart 2015). With the drought increasing the energy required to pump water, this share is bound to increase. California is at a crossroads; either it can double down on non-renewable energy sources, most notably natural gas, in an attempt to merely delay the drought's impact on the energy sector, or it can work now to ensure that California's water needs can be supported by a long-term, sustainable power infrastructure relying on renewable energy.

Exactly what percentage of power used by water utilities derives from renewable sources is unclear. Water utilities in California are not currently asked to disclose what types of energy powers their processes, which makes it difficult to understand the extent to which California's water sector relies on fossil fuels and how difficult transitioning to renewable energy sources might be (UCS 2015). Sonoma County's successful, aggressive plant to use only renewable energy to power their water systems might provide a roadmap for the rest of the state.

In light of a new @UCSUSA report, #drought ridden California looks to #RenewableEnergy to provide for the increasing power needs of the #water sector

Union of Concerned Scientists (UCS). "California's Water Sector Could Benefit From Clean Energy Investments". April 6, 2015. http://www.ucsusa.org/news/press_release/california-water-sector-0487#.VS18dfnIji_

Hart, Angela. "Sonoma County Water Agency hits clean energy goal". The Press Democrat. April 6, 2015. http://www.pressdemocrat.com/news/3762101-181/sonoma-county-water-agency-hits?page=0

Major Energy Leaders Look to Lift United States Embargo on Oil Exports

by Trevor Smith

Senator Lisa Murkowski, an Alaskan Republican and chairwoman of the United States Senate Energy and Natural Resources Committee, promised this week to attempt to lift the US embargo on exporting oil. Many energy leaders gathered at IHS Energy CERAWeek, the global energy conference at which Murkowski delivered the speech, heavily supported the move. The removal of the ban would help to stabilize crude oil prices, which have recently started rising again after plummeting in 2014 (Schaffler 2015). The move comes as US oil production is at a nearly 40 year high.

Murkowski argued that the embargo amounts to the US placing a competitive disadvantage on itself when it comes to the oil market. She noted that the US is in the process of lifting sanctions on Iranian oil even as "we are keeping sanctions on American oil" (Schaffler 2015). The ban was initially put into place to protect American consumers from price volatility and the possibility of increased domestic gas prices (Plumer 2014).

But Murkowski and other energy executives now claim that the ban harms the American economy overall when considering a broader perspective. Since the ban has prevented the US oil from fully entering the oil export and import market, it has forced the US into a position where it must increase or decrease its production based on foreign oil production (Cattaneo 2015). Although this does somewhat decrease the price of gas for American consumers, it creates volatile conditions for US oil companies, who have found themselves forced to slash thousands of jobs because the oil export ban artificially drives down the price of American oil. Those hoping to lift the ban argue that a more stable industry is worth slightly higher gas prices at the pump.

Cattaneo, Claudia. "Chair of US senate energy committee says Washington needs adapt to new era of low oil prices." The Financial Post. April 20, 2015. http://business.financialpost.com/news/energy/chair-of-u-s-senate-energy-commitee-says-washington-needs-to-speed-up-regulatory-approvals-and-lift-oil-export-ban?__lsa=6f7a-51fd

Plumer, Brad. "US oil exports have been banned for 40 years. Is it time for that to change?" The Washington Post. January 8, 2014.
http://www.washingtonpost.com/blogs/wonkblog/wp/2014/01/08/u-s-oil-exports-have-been-banned-for-40-years-is-it-time-for-that-to-change/

Schaffler, Rhonda. "Oil Export Ban Should be Lifted by the US, Energy Leaders Say." The Street. April 20, 2015. http://www.thestreet.com/story/13118779/1/oil-export-ban-should-be-lifted-by-the-us-energy-leaders-say.html

New European Energy Union Hopes to Loosen Russia's Grip on Its Energy Market
by Trevor Smith

The European Commission looked to further loosen the European Union's energy reliance on Russia as it presented its first proposals for a European 'Energy Union.' The idea of such a Union had been floated around by the Commission since the beginning of the year, although it was unclear how the Commission saw the Energy Union's role in interacting with countries outside of the European Union. The proposal is a strong move toward regional energy independence for the European Union, uniting the 28 energy markets of the European Union member states into one large market which will be able to handle fluctuations in energy needs internally (Traynor and Neslen 2015).

Despite initial speculation that Germany would attempt to use its influence in the Energy Commission to increase European dependence on Russian natural gas (Conca), the majority of participants in the Commission have gathered behind Poland to push back against Russian control over European Union energy. The European Union currently relies on Russia for about a third of its oil and gas. But because Russia currently makes oil deals with each member state individually, it is able to charge different countries different prices for the same oil, essentially using the process as a political bargaining chip. Russian oil prices currently vary as much as 20-30% across the European Union (Yablokova 2015). By combining into one large European Energy Union, European countries will be able to pose a stronger, more uniform front when negotiating energy deals.

The proposal has been well received by the member states of the European Union so far, who still have to approve the plan before it can be implemented. The proposal has been altered somewhat in order for member states to maintain a semblance of energy autonomy, allowing for EU states to make bilateral deals with non-EU states under certain conditions. Even still, the Energy Union should still offer the European Union more energy security and control of its energy prices (EurActiv 2015).

Conca, James. "The Role of the New European Energy Union." Forbes. February 9, 2015. http://www.forbes.com/sites/jamesconca/2015/02/09/the-role-of-the-new-european-energy-union/

EurActiv. "Leaders broadly endorse 'Energy Union' plans, leave details to later." Euractiv. March 20, 2015. http://www.euractiv.com/sections/energy/eu-leaders-back-commissions-energy-union-plans-313093

Tryanor, Ian and Neslen, Arthur. "Ambitious EU blueprint for energy union to loosen Russian grip on gas." February 24, 2015. http://www.theguardian.com/world/2015/feb/24/eu-blueprint-energy-union-russian-gas-gazprom-maros-sefcovic

Yablokova, Alina. "New EU Energy Union Aims to Break Dependence on Russia." International Policy Digest. March 7, 2015. http://www.internationalpolicydigest.org/2015/03/07/new-eu-energy-union-aims-to-break-dependence-on-russia/

Will Sage Grouses Stop Green Energy Development?

by Abigail Wang

The debate over the fate of the sage grouse, a bird known for its elaborate mating display, has been ongoing for over 15 years. Due to the species' alarming population decline in the United States, the development of green energy in the West has slowed considerably. Environmental groups have thus become divided in answering the difficult question of wind and solar energy expansion versus wildlife protection.

The sage grouse has experienced a 45-80% decline in population since the 1800's and are currently found on federally owned lands in only 11 western states including Wyoming, Utah, California, Idaho, and Colorado. As the birds' habitats continue to shrink, the United States Fish and Wildlife Service has received numerous petitions since 1999 to list the bird as a threatened or endangered species. A 2010 settlement agreement decreed that the FWS must determine whether or not it would list the sage grouse under the Endangered Species Act (ESA) or remove it from the candidate list by September 2015.

Due to political pressure, the FWS recently declared that the sage grouse would be listed as 'threatened' and not 'endangered'. This caused an outcry among several conservationist and environmentalist groups who argue adamantly that the sage grouse should be listed as endangered. Listed as 'threatened', the sage grouse will not receive protections including the prohibition of killing, harassing, or destroying the birds' habitat. In fact, three environmentalist groups—WildEarth Guardians, Center for Biological Diversity, and Western Watersheds Projects—filed suits against the federal government for not doing enough to protect the bird.

Wind energy facilities like turbines are likely to negatively impact sage-grouse population as well. The birds tend to avoid areas near the structures, so building turbines too close to their breeding sites would drive them away. This doesn't factor in the roads and maintenance facilities that would also be built over habitat areas to run the facilities.

Nonetheless, the current 'threatened' listing of the sage grouse could further inhibit the development and progress of green energy in the West. Even prior to the decision, state protection programs like the Wyoming Core Area Strategy, limited development in the main sage-grouse areas to 5% disturbance. Programs like these have taken a toll on state economies, delaying and preventing energy projects.

The economic impact is a major concern because the Western states that make up the sage grouse's habitats account for 27% of the total energy produced in the United States. Other environmental organizations have attempted to take the middle road between green industries and sage-grouse enthusiasts, insisting that green energy facilities can successfully coexist with the bird as long as they are built farther away from breeding sites. Whether or not the development of green energy suffers at the wings of the sage grouse remains to be seen.

Stoellinger, Temple. "Implications of a Greater Sage-Grouse Listing on Western Energy Development." National Agricultural and Rural Development Policy Center. June 2014.
http://www.nardep.info/uploads/Brief33_ImplicationsListingSageGrouse.pdf

LeBeau, Chad, Jeff Fruhwirth, and J.R. Boehrs. "Analysis of the Overlap between Priority Greater Sage-Grouse Habitats and Existing and Potential Energy Development Across the West." Western Values Project. October 2014: 1-26.
http://westernvaluesproject.org/wp-content/uploads/2014/10/Greater-Sage-Grouse-Priority-Habitats-and-Energy-Development.pdf

Brown, Matthew and Mead Gruver. "Fat of The Struggling Greater Sage Grouse Shaping Energy Development In US West." Huffington Post: 9 December 2014.
http://www.huffingtonpost.com/2014/12/04/greater-sage-grouse-us-energy_n_6269806.html

Strickland, Dale. 2014. "Greater Sage-Grouse and Energy Development" [Powerpoint slides]. Retrieved from http://wyia.org/wp-content/uploads/2014/12/WEST-WIA-sage-grouse-energy-presentation-Chad.pdf.

Robertson, Erin. 2014. "Sage-Grouse and Wind Energy: The West is Where the Wind Blows." http://rockymountainwild.org/_site/wp-content/uploads/Sage-Grouse-and-Wind-Energy.pdf

"Environmentalists ask court for more protection on sage grouse, found only in Colorado, Utah." Fox Business 20 January 2015.
http://www.foxbusiness.com/markets/2015/01/20/environmentalists-ask-court-for-more-protection-on-sage-grouse-found-only-in/

Protecting Alaskan Wilderness At What Cost?
by Abigail Wang

As President Obama finishes his last term, he's rolling full steam ahead with his environmental and energy policies. In a move that left environmentalists, oil companies, and politicians upset, the president announced the Interior Department's plans to prevent future oil and gas production in major parts of Alaska, but support development along the East Coast.

The Obama administration wants to designate 12.28 million acres of the Arctic National Wildlife Refuge (ANWR), including the coastal plains in Alaska, as "Wilderness". Wilderness is the highest level of protection available for public lands; it prohibits mining, drilling, roads, vehicles, and the establishment of permanent structures in select areas. Over seven million acres are currently managed as wilderness because of the National Interests Lands Conservation Act of 1980, but more than 60% of the ANWR is not listed as such.

The plan addresses several points including the protection of wildlife populations and habitats, opportunities for fish and wildlife recreation, and support for the needs of local inhabitants. It emphasizes the importance of monitoring wildlife, habitats, and the public use of the land in order to respond to the effects global warming.

Alaskan Republicans are infuriated with Obama's decision, claiming that if the plans were enacted, the state's economy would greatly suffer. The state is already facing a $3.5 billion budget shortfall, and oil producers are likely to stop pursuing development if it's no longer economical. Alaskan Senator Lisa Murkowski, the chairman of the Senate Energy and Natural Resources Committee, vowed to try everything possible to block Obama's plan. Likewise, Governor Bill Walker stated he would increase the oil exploration and production on state-owned lands in order to make up for federal restrictions.

This isn't the first time the president and Congress have fought over the Arctic. The first showdown occurred in 1995 when President Bill Clinton vetoed legislation passed by Congress that would have approved exploration and production on Alaska's coastal plain. Fifteen years later Senators Murkowski and Ted Stevens tried to open a non-wilderness portion of the ANWR to development but supporting votes fell short.

While major environmental leaders like Patagonia and Wilderness Society hail the proposal as a good step in the long-term vision of the region's future, local advocacy groups on the

East Coast are less enthusiastic. The measures may protect the habitat of threatened polar bears, porcupine caribous, and musk oxen, but don't offer the same for the creatures in the Atlantic.

The administration is planning to open a long stretch of the East Coast waters, including the southeast stretch of the Atlantic Seaboard, to oil and gas drilling. Environmentalists are concerned that opening Atlantic waters is going put the coasts of Virginia, the Carolinas, and Georgia at risk for a disaster reminiscent of the 2010 BP oil spill in the Gulf Coast. The administration will have to compromise to see if it can appease all parties while pushing its agenda through.

Barrett, Ted. "Alaska Delegation Rips White House Decision." CNN: 26 January 2015. http://www.cnn.com/2015/01/26/politics/alaska-delegation-white-house-anwr/

Berman, Dan and Ben Geman. "Obama Proposes to Declare ANWR Area as Wilderness." The National Journal: 25 January 2015. http://www.nationaljournal.com/white-house/obama-proposes-to-declare-anwr-area-as-wilderness-20150125

Bradner, Eric. "Obama's Alaska Move Triggers Fight." CNN: 26 January 2015. http://www.cnn.com/2015/01/25/politics/obama-alaska-energy-fight/index.html

Davenport, Coral. "Obama's Plan: Allow Drilling in Atlantic, but Limit It in Arctic." New York Times: 27 January 2015. http://www.nytimes.com/2015/01/28/us/obama-plan-calls-for-oil-and-gas-drilling-in-the-atlantic.html?_r=0

Eilperin, Juliet. "Obama Administration to Propose New Wilderness Protections in Arctic Refuge – Alaska Republicans Declare War." The Washington Post: 26 January 2015. http://www.washingtonpost.com/news/energy-environment/wp/2015/01/25/obama-administration-to-propose-new-wilderness-protections-in-arctic-refuge-alaska-republicans-declare-war/

"Obama Administration Moves to Protect Arctic National Wildlife Refuge." US Department of the Interior: 25 January 2015. http://www.doi.gov/news/pressreleases/obama-administration-moves-to-protect-arctic-national-wildlife-refuge.cfm

Sanders, Sam. "Obama Proposes New Protections For Arctic National Wildlife Refuge." NPR: 25 January 2015. http://www.npr.org/blogs/thetwo-way/2015/01/25/379795695/obama-proposes-new-protections-for-arctic-national-wildlife-refuge

The Unseen Problems with Nevada's Air Quality
by Abigail Wang

For nearly twenty years Nevada, along with parts of Arizona, has been a hot topic of the debate between public health and economic development. The issue has resurfaced again as Brenda Buck and Rodney Metcalf, geologists and professors of geoscience at University of Nevada, push the Nevada Department of Health to implement more protective measures in areas filled with asbestos.

Asbestos is a naturally occurring carcinogenic fiber that damages the lungs. Deposits of asbestos minerals are fairly common and particularly rich veins have been mined for commercial use. The fibers travel easily through and air and if inhaled, even in modest amounts, can embed themselves into the lungs and cause mesothelioma and other respiratory diseases. Mesothelioma, a type of lung cancer, can only be caused by the exposure to asbestos, and it normally takes 30 or more years to recognize symptoms.

After analyzing the dunes and outcroppings of southern Nevada, Buck and Metcalf discovered the landscape was filled with asbestos. Even though asbestos is found in most parts of the country, natural erosion and commercial development in Nevada send these fibers into the wind. After investigating asbestos-related diseases, the geologists worked with epidemiologist Francine Baumann of University of Hawai'i to present a report that found a pattern of mesothelioma in the affected areas of Nevada. However, when the Nevada Department of Health learned of the report, it forced Baumann to withdraw the presentation of her findings and revoked her access to the state cancer registry (Blum 2015). Critics, like state epidemiologist Dr. Ihsan Azzam, argue that the scientists are propagating unnecessary fear about asbestos and the presence of the mineral does not necessarily equate to human exposure and negative health consequences.

Continued research on the fiber has shown that natural asbestos is found from Lake Mead to McCullough Range, in Boulder City, Henderson, Las Vegas, and even in parts of Arizona. Those who have been diagnosed with mesothelioma in these areas are exposed to the same asbestos fibers found in Libby, Montana, where the population has suffered decades of asbestos-related illnesses. Reports have shown that those in the affected Nevada and Arizona areas are as young as 15, suggesting that this unusual rise in cases is linked to environmental exposure to the carcinogenic fibers.

Due to the recent findings, Nevada's Department of Transportation has delayed construction of a $490 million highway project, and a federal interstate highway connecting Las Vegas and Phoenix may also be put on hold.

Metcalf and Buck have championed for more protective measures, like wearing facemasks in areas of high asbestos concentration and limiting outdoor activities on windy days. Even though representatives of the Nevada Department of Health argue that there is little cause for concern, the department has increased monitoring of the airborne fibers as a precaution.

Blum, Deborah. "In Nevada, a Controversy in the Wind." NY Times: 9 February 2015. http://www.nytimes.com/2015/02/10/science/a-controversy-in-the-wind.html?ref=science

International Association for the Study of Lung Cancer. "Mesothelioma in Southern Nevada Likely Result of Asbestos in Environment." ScienceDaily: 10 February 2015. http://www.sciencedaily.com/releases/2015/02/150210155940.htm

Kyser, Heidi. "Waiting to Inhale." Desert Companion: February 2015. http://www.desertcompanion.com/article.cfm?ArticleID=1067

Morell, Casey and Natalie Cullen. "Health Concerns and Highway Expansion Converge in Boulder City." KNPR News: 11 February 2015. http://knprnews.org/post/health-concerns-and-highway-expansion-converge-boulder-city

A Green Makeover for France
by Abigail Wang

Bon travail, France! In an effort to go green as its European neighbors expand their renewable energy industries, the French government passed legislation that requires the roofs of new commercial buildings to be at least partially covered by either solar panels or plants.

Green roofs have recently gained popularity, most notably in Toronto, Canada and Basel, Switzerland, where similar mandatory green roof laws were passed. Not only is it cost-effective—cities could save hundreds of millions of dollars in energy cost—but rooftop vegetation also solves a lot of energy and environmental problems. Green roofs save and generate electricity, insulate buildings, reducing the need for heating and air conditioning, and can temper the urban heat island effect. They also help reduce runoff by retaining rainwater, improve air quality by absorbing pollutants, and provide urban habitats for birds.

So why aren't all countries jumping on the bandwagon? Green roofs are much more difficult and expensive to construct, and it's nearly impossible to greenify a roof of a building that simply wasn't designed for it. However, similar to rooftop solar panels, long-term savings cover the significant upfront cost and upkeep. Green rooftops also last two to three times longer than conventional roofs.

Environmental groups lobbied for the law to apply to the construction of all new buildings and wanted total rooftop coverage with greenery. Even though their demands weren't fully met, the fact that the legislation passed signals the French government's willingness to engage with more renewable energy policies.

France is primarily dependent on nuclear energy—in 2012 alone, over 80% of its electricity came from this source—but is trying to take steps to be greener. Two wind turbines were installed in the Eiffel Tower in late February, and streets in Paris boasted camouflaged wind turbines that are shaped like trees in March. After lagging behind its neighbors, France is making its way up the green energy initiative ladder.

Ackerman, Evan. "New Commercial Buildings in France Must Get Green Roofs or Solar Panels." The Green Energy Blog: 26 March 2015.
http://thegreenenergyblog.com/uncategorized/new-commercial-buildings-in-france-must-get-green-roofs-or-solar-panels

Appleyard, David. "France Mandates Solar Roofs for Commercial Zones." RenewableEnergyWorld.com: 30 March 2015.
http://www.renewableenergyworld.com/rea/news/article/2015/03/france-mandates-solar-roofs-for-commercial-zones

Cheeseman, Gina-Marie. "Bonne Nouvelle! French Law Mandates Green Roofs, Solar Panels." TriplePundit: 25 March 2015.
http://www.triplepundit.com/2015/03/bonne-nouvelle-french-law-mandates-green-roofs/

Koch, Wendy. "Pictures: Green Roofs Get Lift As France Makes Them de Rigueur." National Geographic: 28 March 2015.
http://news.nationalgeographic.com/energy/2015/03/150328-green-roofs-france-buildings-energy/

Crikey! Australia Could Reach 100% Renewables

by Abigail Wang

Australia might be able to add something to its long list of things it's known for. Everyone is familiar with the Great Barrier Reef and the Sydney Opera House, but soon people might also know the continent down under as one of the foremost leaders in renewable energy. According to a report from the Australian National University's Center for Climate Economics and Policy, Australia has the potential to reach 100% renewables and zero net emissions by 2050.

Thanks to the decrease in costs of carbon-free technologies, like wind and solar, Australia is well poised to reduce its emissions. The report claims that Australia could source 100% of its power from renewables by 2050 without depressing economic growth or incurring great adjustment costs. The continent is already utilizing wind turbines, solar panels, and concentrating solar thermal power. As long as the government sets clear and stable policies to support investments in these and other renewables, there is a good chance Australia could achieve this ambitious goal.

Critics of this goal argue that this shift could cause major problems. There will be a massive fight over the early retirement of significant existing power generation assets, and some think that decarbonization will cost too much. However, the report alleges that previous work has overestimated the costs associated with reducing pollution, and that the costs of solar and wind energy have fallen faster than anticipated. Also, any surplus or deficit problems could be addressed by molten salt storage associated with solar thermal plants, biomass-fired generators, and existing hydropower.

A major obstacle holding this goal back is political stability and vision. The Australian government is notoriously known for favoring pro-coal agendas, and new investments in large-scale renewables are practically stalled. If Australia stops exporting coal and natural gas, it's unclear what the economic consequences will be. The goal is certainly achievable as long as the government, companies, and individuals cooperate.

"100% Renewable Energy Powered Australia Possible by 2050." Energy Matters: 22 April 2015. http://www.energymatters.com.au/renewable-news/renewables-anu-wwf-em4784/

Hill, Joshua S. "Australia Could Reach 100% Renewables by 2050, as Experts Recommend 30% Emissions Cut by 2025." 22 April 2015. http://cleantechnica.com/2015/04/22/australia-reach-100-renewables-2050-experts-recommend-30-cut-2025/

Murphy, Katharine. "Australia could source 100% of power from renewables by 2050, report finds." The Guardian: 20 April 2015. http://www.theguardian.com/environment/2015/apr/20/australia-could-source-100-of-power-from-renewables-by-2050-report-finds

ENERGY FINANCE AND ECONOMICS

It takes money to get energy. Where does that money come from? Should governments subsidize energy production? Are investments in renewable energy prudent? There are many points of view.

Crowdfunding Goes Where No Government Has Gone Before: Caring about the Environment
by Hannah Brown

Well, maybe that's a slight exaggeration. Governments and organizations across the globe have made efforts to ameliorate, or at least, modify, the consequences of climate change but not always at the pace that scientists urge. Reports such as those conducted by the Intergovernmental Panel on Climate Change (IPCC) show us that the earth is in a dire state, that climate change has been definitively caused through human action, and that intervention is of the utmost importance. However, while this might be a red flag for most, places like Australia where the Climate Commission was abolished and only resurrected through community fundraising, are examples of a misdirection of attention. But, as Australia's Climate Commission shows, where large organizing bodies are failing, individuals and innovators are rising.

Cue the armchair activists and their armchair investing. Crowdfunding sites, such as Indiegogo and Kickstarter, have become unique platforms for people who have potentially successful environment minded ideas but lack funding. There's even a handout on how to do it (Salon.com).

The types of projects are varied and intriguing. Established researchers such as Ralph Keeling, the heir to CO2 research in the US, used crowdfunding when his grants from the National Science Fund were cut. Other researchers proposed more far out, but not entirely implausible projects, such as the Solar

Roadways. This project, which had been publicly funded by the Department of Transportation, raised more than $2.2 million dollars to continue their work developing roadways created out of solar panels and LED lights.

There are projects that find a happy medium—where they attract the individual, and promise moderate, but successful, change. A quick Google search will lead to a number of potential projects—from supporting rainforests in tropical Queensland, to helping fund community gardens to building communities through solar campaigns. There is a project that everyone can get behind.

However, some have pointed out that while there are benefits to crowdfunding, there are drawbacks as well. Questions like: how can the public know which ones are truly worth supporting, and which ones just have a good video and media presence? Additionally, journalists such as Alice Bell, ask if it takes the responsibility of solving climate change issues unduly away from of the government, letting them off the hook when it should be one of their greatest priorities.

Regardless of the problems, crowdfunding does engage in the individual in climate change more than most other options as it can spur curiosity and lead to education. It also gives people an opportunity to make an impact and realize their stance and value in fighting climate change.

Leonard, Andrew. "Climate change madness: The fate of the planet now depends on Kickstarter." Salon.com. January 10, 2014.

Lacey, Stephen. "Would Solar Roadways Work? A Government Engineer Discusses the Controversial Technology." Greentechmedia.com. August 29, 2014.

Von Ritter, Konrad and Black-Layne, Diann. "Climate Change: A new source of finance for climate action at the local level?" European Capacity Building Initiative. May 2013.

North End Organic Nursery. "Bring the Garden back to Garden City." Kickstarter.com

Intergovernmental Panel on Climate Change. "Findings of the IPCC Fourth Assessment Report: Climate Change Science" UCSUSA.com.

Flannery, Tim. "Climate Council raises $1m through its Obama-style fundraising drive" The Guardian. October 5, 2013.

Citi Deploys $100 Billion on Clean Energy
by Nour Bundogji

Citi, the leading global bank, announces a sustainability strategy that should last for over ten years— $100 billion for lending, investing, and facilitating sustainability solutions. This eye-popping financial commitment is part of Citi's five-year plan that was launched by CEO Michael Corbat in New York last week. "It includes three strategic priorities that align the company's corporate and sustainability strategies: combating climate change, championing sustainable cities, and promoting social progress, including universal human rights" reports Joel Makower from Greenbiz.com who sat down with Val Smith, Citi's director of corporate sustainability. Furthermore, Corbrat informed Mary Lubber at Forbes.com that this strategy will include "financing for large renewable-energy projects such as municipal infrastructure to reduce water waste; assistance for clients to address environmental risks; and an 80% absolute greenhouse gas reduction target."

Citi's commitment to sustainability and clean energy is not new. In 2007, Citi made its first financial commitment of $50 billion for a ten-year pledge to "invest in and finance projects that reduce global carbon emissions" says Makower. However, Citi met its financial goal three years earlier than expected. The new $100 billion goal "builds on the learnings that we accumulated during the first $50 billion. When we knew that we were going to hit the $50 billion goal three years early, we did an assessment across our different businesses. We wanted to expand the scope of the $100 billion goal so that we could capture a lot of other activities that are very important to people, that are very important to cities, that our clients are deeply engaged in," Smith told Makower.

Nonetheless, Citi was one the first major banks to set reduction goals for energy, waste, and water. For instance, in 2003 it was one of the ten global banks to sign the Equator Principles, which is a risk management framework for financial institutions to help them determine, assess, and manage environmental and social risks.

Bloomberg New Energy Finance recently released the clean energy investment numbers for 2014. Although clean energy investment increased 16% ($310 billion, an all-time high), "it's just not enough," says Lubber.

Regardless, "Citi is showing that investing in clean energy is smart business, and that – with a bit of ambition and commitment – it can be done right now. It's a clear market signal that should resonate with the industry," claims Lubber. Although Citi has a long road ahead, this $100 billion commitment is a critical step in the right direction showing what corporate leaders are capable of.

Lubber, M. Citi's $100 Billion Downpayment On Clean Energy Future. Forbes.com. Feb. 20, 2015.
http://www.forbes.com/sites/mindylubber/2015/02/20/citis-100-billion-downpayment-on-clean-energy-future/2/
Makower, J. Inside Citi's plan to deploy $100 billion for cities, renewables, climate. Greenbiz.com. Feb. 18, 2015.
http://www.greenbiz.com/article/inside-citis-plan-deploy-100-billion-cities-renewables-climate-solutions

Bloomberg Philanthropies Building Greener Cities

by Jessie Capper

Although many know Michael Bloomberg as the past Mayor of New York City, holding the position for three consecutive terms from 2001–2004, his company Bloomberg Philanthropies demonstrates that he is much more than an active politician in the US government. Bloomberg Philanthropies primary mission is to help the largest number of people live the best, and healthiest lives possible. Through harnessing his entrepreneurial spirit, discovering viable solutions, using data to assess financial reasonability, advocacy, and partnerships with other organizations (both private and public), Bloomberg Philanthropies work to address a multitude of issues facing our world—starting within cities (Bloomberg Philanthropies). Most interesting is Bloomberg Philanthropies' most recent partnership with the Heising-Simons family to begin a Clean Energy Initiative, supporting city initiatives to develop a cleaner, more sustainable energy system within local power grids (Green Tech Media). Due to the out-of-date, harmful practices of our current cities, I see this partnership as a promising start towards achieving Bloomberg Philanthropies' goal, and making clean energy the "norm."

Unfortunately, a majority of cities' power is dependent on the respective local power plants. These power plants account for approximately 38% of the United States' greenhouse gas emissions. Fortunately, alternative energy resources such as solar, wind, and LED, are proving to be more cost-efficient and sustainable substitutes for our current malpractices. The Clean Energy Initiative will encourage cities to comply by the Environmental Protection Agency's Clean Power Plan, thereby adjusting cities' current models for attaining power to a set of guidelines that reduce our carbon pollution. Our world today is abuzz is with research, promotion, and development of technologies that can help achieve the EPA's goal of increasing renewable energy production three times by 2025. Personally, one big area of concern, however, is the dependability of these technologies and alternative energy sources. Although it is exciting to see increasing support and dependence on this new sector, I believe we must consider their respective reliability in comparison to our current power plants. With their work on previous projects, however, I have faith Bloomberg Philanthropies and the Heising-Simons family will address this concern while assisting city and state officials in developing the most environmentally friendly power strategies.

Renewables Biz (http://www.renewablesbiz.com/article/15/01/bloomberg-philanthropies-and-partners-launch-new-clean-energy-initiative-limit-carbon-pollution-power-plants-and-s)

Bloomberg Philanthropies (http://www.bloomberg.org/about/our-approach/)

Green Tech Media (http://www.greentechmedia.com/articles/read/Bloomberg-Heising-Simons-Donate-48M-to-Cut-Carbon-and-Deploy-Clean-Energy)

TAGS: Michael Bloomberg, Bloomberg Philanthropies, Heising-Simons Foundation, EPA, Clean Energy Initiative, Energy efficient technology

Meghan@bloomberg.org

Making Solar Energy an Option for All Families

by Jessie Capper

The United States' federal government offers an energy assistance plan to low-income families who would not otherwise be able to afford their utility bill. Although this program is largely beneficial to these communities, it is not advantageous for our current environmental conditions. Unfortunately, the energy assistance provided by the government does not provide the recipient the opportunity to choose the source of their energy; in other words, the Low Income Home Energy Assistance Program formed by the government only supplies energy from the local utility companies. These companies supply a majority of their energy from power plants, which account for roughly 38% of the United States' greenhouse gas emissions (Bloomberg Philanthropies). Fortunately, the Rural Renewable Energy Alliance is changing this, ultimately building a sustainable renewable energy program for families of any financial background (RREAL 1).

The Rural Renewable Energy Alliance has developed the Solar Assistance program, also coined the "Community Solar for Community Action" project, to make renewable energy an option for all individuals—regardless of their respective economic condition (RREAL 2). RREAL offers these families with the option of receiving their energy from local community solar gardens. The Solar Assistance program aims to build a solar garden in these lower-income communities that would ultimately be owned and managed by a community action organization. This local group would organize the distribution of the garden's solar energy to the families currently on or eligible for federal energy assistance (Ayre Jan 30, 2015). Through RREAL's Solar Assistance program, these families would save on rising utility costs from fossil fuels while creating a self-reliant, environmentally beneficial, and socially equitable program.

Unfortunately, RREAL is still finalizing many of the details of their plan, and there is no quantifiable value for the number of families they can properly affect. They hope to benefit no fewer than 40 families; however, this number, and the overall success of the program, is currently unpredictable. Although all the details of the plan are yet to be finalized, RREAL received a grant from the McKnight Foundation in December to support the initial stages of the project, which is set to first be implemented in Minnesota (McKnight Foundation); the Rural Renewable Energy Alliance and the McKnight Foundation are confident that the Solar Assistance program will expand beyond Minnesota, ultimately altering and benefitting the financially isolating renewable energy system we have today.

Ayre, James. "Rural Renewable Energy Alliance — Helping Low-Income Families Switch To Solar." CleanTechnica. January 30, 2015. Accessed February 4, 2015. http://cleantechnica.com/2015/01/30/rural-renewable-energy-alliance-helping-low-income-families-switch-solar/

Bloomberg Philanthropies (http://www.bloomberg.org/press/releases/bloomberg-philanthropies-partners-launch-new-clean-energy-initiative-limit-carbon-pollution-power-plants-spur-clean-energy-investments/).

McKnight Foundation (http://www.pineandlakes.com/news/business/3641739-mcknight-foundation-awards-grant-rreal).

RREAL 1 (http://www.rreal.org/).

RREAL 2 (http://www.rreal.org/solar-assistance/).

Are lower oil prices really a concern for the renewable energy industry?

by Jessie Capper

According to Bloomberg New Energy Finance, funding for wind, solar, biofuels, and other low-carbon energy projects increased 16% in 2014—amounting to $310 billion and signifying the first growth in financial support since 2011 (Downing Jan 9, 2015). The BNEF claims that this increase in funding is a result of a 32% expansion in China's commitment to renewables, including the nation's $19.4 billion recently dedicated to offshore wind projects (Downing Jan 9, 2015). Although this growth in funding is promising for the renewable energy industry, price drops in oil have caused concern that this expanding support will be short-lived (Downing Jan 9, 2015).

Recent trends in solar demand, however, demonstrate otherwise. A solar boom in 2014 resulted in the first global panel shortage since 2006, revealing an increase in interest and investment among renewables despite falling oil prices (Goossens Aug 19, 2014). A two-year slump prior to the solar boom lowered prices of solar power, making the technology a competitive alternative to oil-based energy sources. According to Michael Park, an analyst at Stanford C. Bernstein, there is a staggering price relationship between solar and fossil fuels. Park demonstrates that solar prices are falling rapidly, and will soon "undercut even the cheapest fossil fuels, coal, and natural gas. In the few places oil and solar compete directly, oil doesn't stand a chance" (Randall Jan 30, 2015).

Anticipating an increase in demand, SunPower Corp in San Jose, California is manufacturing at its full capability and has announced plans for a new factory to begin production in 2017, producing at least 700 megawatts a year (Goossens Aug 19, 2014).

Past trends in oil prices show that oil prices will not stay this low forever. Goldman Sachs determined that the almost $1 trillion investments in future oil projects are no longer profitable when oil is under $70 a barrel; therefore, American drillers are idling rigs faster than they have since 1991 (Randall Jan 30, 2015). Expanding demand in the solar industry and the increasing risk in the oil industry all demonstrate that lowering oil prices will not stop the increasing funding and progress in the renewables industry.

Downing, Louise. "Clean Energy Investment Jumps 16%, Shaking Off Oil's Drop." BloombergBusiness. January 9, 2015. Accessed March 12, 2015. http://www.bloomberg.com/news/articles/2015-01-09/clean-energy-investment-jumps-16-on-china-s-support-for-solar

Goossens, Ehren. "Solar Boom Driving First Global Panel Shortage Since 2006." BloombergBusiness. August 19, 2014. Accessed March 12, 2015. http://www.bloomberg.com/news/articles/2014-08-18/solar-boom-driving-first-global-panel-shortage-since-2006

Randall, Tom. "Seven Reasons Cheap Oil Can't Stop Solar Power Anymore." BloombergBusiness. January 30, 2015. Accessed March 12, 2015. http://www.bloomberg.com/news/articles/2015-01-30/seven-reasons-cheap-oil-can-t-stop-renewables-now

The Cost of Solar Power
by Alex Elder

Solar panels have long been touted as a simple source of renewable energy. They have even become widely available to average energy consumers; a homeowner can simply attach solar panels to their roof and gain access to solar power energy. Utilization of solar panels has increased greatly over the past several years, with one rooftop solar system being installed every four minutes in 2013 (Than 2013). However, despite the availability of solar panels and their ease of use, some people have raised concerns about their impact on the energy market.

This new source of energy is not good for electric companies because as more and more solar panels are installed, the smaller their dividends get. Although this may seem like a positive consequence of increased reliance on renewable technology, electric companies disagree. They feel as though people who receive subsidized electricity bills through their use of solar panels are not paying their fair share. Although solar panels are a good source of electricity, consumers still rely on the infrastructure originally built by electric companies to connect households to the electric grid. Thus, users of solar panels are depending upon this groundwork without paying the cost, or so electric companies believe.

As a result of this perceived imbalanced cost, electric companies have tried to scale back the discounts solar panel users receive. Electric companies all over the country are petitioning local governments to reduce the government-provided rebates from rooftop solar systems. The cost of utilities, these companies argue, is not simply for the energy consumed but also for access to the electric grid. David Owens, executive vice president at the Edison Electric Institute stated that "it's not about lost revenue. We want to make sure the grid is maintained, that it can be enhanced." However, it is unclear whether or not the electric companies' backlash against residential solar energy is truly about grid access or is simply a cover for an effort to regain lost profits.

Than, Ker. As Solar Power Grows, Dispute Flares Over US Utility Bills. The National Geographic. Published December 24, 2013. - http://news.nationalgeographic.com/ news/energy/2013/12/131226-utilities-dispute-net-metering-for-solar/

Kind, Peter. Disruptive challenges: financial implications and strategic responses to a changing retail electric business. Edison Electric Institute, 2013.

Carley, Sanya. "State renewable energy electricity policies: An empirical evaluation of effectiveness." Energy Policy 37.8 (2009): 3071-3081.

Price Premiums for Homes with Rooftop Solar
by Liza Farr

The Lawrence Berkeley National Laboratory recently released a study on solar PV systems housing premiums using the largest data set to date (Hoen *et al.* 2014). PV costs have dropped drastically in the past several years, and innovative financing options such as power purchase agreements and solar leasing have made solar an increasingly popular addition to residential homes. The new study, using a hedonic pricing model, reveals an average of around $4 per watt increase in housing price for a PV home over a non-PV home. This approximates to about a 0.92% increase in value for each kilowatt installed. There also appears to be a "green cachet," meaning buyers are willing to pay a base amount for having PV at all, with incremental increases in willingness to pay with increases in the size of the system. Unfortunately, although to be expected, there is a clear decrease in price premiums as the systems age. Since one of the biggest drawbacks of solar panels are the high input costs, this return in the form of a housing price premium could convince many homeowners to make the purchase (Hoen *et al.* 2014).

This has broad implications for solar stakeholders, housing developers, and homeowners. Stakeholders now have defensible, concrete evidence promoting the economic viability of going solar, and the subsequent viability of solar companies as a good investment. Housing developers can increase the price of their homes if they use solar, as well as the boost they will get in their public image. Developers can also advertise homes with large south facing roofs as an asset for homeowners to put their own solar panels on, and receive the price premium. With this information, homeowners will increase their deployment of rooftop solar, perhaps in place of other remodeling that can be just as expensive, with less of a payback. The newly popular leasing option now appears less appealing, as Fannie Mae has indicated leased systems cannot be included in the appraised value of the property (Hoen *et al.* 2014). Undoubtedly, this is good news for solar power, and we can look forward to accelerated deployment of residential PV, more jobs in the solar industry, and reduced greenhouse gas emissions as a result.

Hoen, B., S. Adomatis, T. Jackson, J. Graff-Zivin, M. Thayer, G.T. Klise, and R. Wiser. 2014. Selling into the Sun: Price Premium Analysis of a Multi-State Dataset of Solar Homes. Berkeley, CA: Lawrence Berkeley National Laboratory.
Fact Sheet for Selling into the Sun (http://emp.lbl.gov/sites/all/files/selling-into-the-sun-fullreport-factsheet-jan12.pdf)

Google Makes New Clean Energy Purchases After Abandoning R&D Efforts
by Liza Farr

During the last few years, Google has had a roller coaster ride with clean energy. In 2007, they started their RE<C initiative, which aimed to develop renewable energy sources that would be cheaper than coal, through research and development (Romm, Dec 4, 2014). In 2011, however, Google stopped this program completely, with the reasoning that they determined this effort would not actually reverse climate change, or make

new renewable energy cheaper than existing coal. Critics argued that both of these goals are widely considered impossible, and chastised the company for ending the R&D program. Likely, Google realized with plummeting global renewable energy prices, there was more money to be made in renewable energy deployment than in research and development (Romm, Dec 4, 2014).

By 2014, Google was powered 34% by renewable energy, and rising (Berniker, Feb 16, 2014). Many Silicon Valley tech companies have invested heavily in renewable energy, partly due to their high costs on data centers and infrastructure spending, and partly to create an image that they are environmentally and cost sensitive (Berniker, Feb 16, 2014). In January of 2015, Google continued to ramp up their purchasing of renewable energy. The company backed a 104 megawatt (MW) solar power plant in Utah, and signed a 1,040 MW wind energy contract for their $600 million data center in the Netherlands, powered entirely by renewable energy. Most prominently, this year Google also signed a long-term power purchase agreement with NextEra Energy Resources for the energy generated by the iconic Altamont Pass wind farm (Hill, Feb 13, 2015). Company officials say they will be using this energy to power their north bayshore campus in Mountain View, which makes the first time they will be using renewable energy to power one of their offices. The new wind turbines will generate 43 MW of clean energy, and Google plans to purchase 24 new turbines for the farm in order to generate as much power as the campus uses in a year, making it completely clean powered. Altamont Pass was created in the 1980s as the testing site for large scale wind power in the United States, making it an iconic and important purchase for Google (Hill, Feb 13, 2015). Undoubtedly Google is making the statement that they will be leading the Silicon Valley pack in their green energy investment and use, appealing to the many environmentalists across the nation, and appealing to their shareholders who like to see the company cutting costs where possible. Research and development efforts are still necessary to bring clean energy costs down even further, but clean energy companies and environmentalists alike are still happy to see the massive purchases the company is making to offset their similarly massive energy use and carbon footprint.

Berniker, Mark. "Google makes huge investment in clean energy." Feb 16, 2014. [http://www.cnbc.com/id/101417698#]

Hill, Joshua S. "Google signs on to repower iconic bay area wind farm." Feb 13, 2015. [http://cleantechnica.com/2015/02/13/google-signs-repower-iconic-bay-area-wind-farm/]

Romm, Joe. "The strange thing about Google's decision to stop renewable energy research." Dec 4, 2014.
[http://thinkprogress.org/climate/2014/12/04/3597629/google-engineers-climate-change/]

Rise in Global Renewable Energy Investment Reveals Market Trends
by Liza Farr

A recent report by the United Nations called "Global Trends in Renewable Energy Investment" found exciting results for the renewables industry (Hruska, Apr 2, 2015). After two years of declining investment, the dollars spent on renewables shot up again in 2014 (Hruska, Apr 2, 2015). The investment of $270.2 billion indicates an almost 17% increase since 2013 (Hruska, Apr 2, 2015). There are many explanations for this, including a surge in solar power investment in Japan and China, and new wind turbine construction for off-shore plants in the North Sea (Mooney, Mar 31, 2015). China alone invested 37% more in renewables than during the previous year (Hruska, Apr 2, 2015). Brazil, India, South Africa, Chile, Mexico, Kenya, and Turkey all invested over a billion dollars each in green electricity generation (Hruska, Apr 2, 2015). A large portion of the investment was in wind and solar: $99.5 billion and $149.6 billion respectively (Mooney, March 31, 2015). The other striking element of the report's findings is that developing nations have caught up with their developed counterparts in renewable energy investment (Hruska, Apr 2, 2015). They are investing more in renewables than in other types of power generation, and nearly half of the total investment in 2014 came from developing nations (Mooney, March 31, 2015). Since these countries have less advanced infrastructure, they benefit from the small-scale grids, local production, and lower upkeep costs of solar and wind power (Hruska, Apr 2, 2015). Additionally, renewables are more reliable than traditional fuels that use infrastructure that requires maintenance and a central supplier because there is often civil and economic unrest in these countries (Hruska, Apr 2, 2015). Additionally, oil and gas prices fluctuate tremendously, whereas renewable costs constantly declined (Hruska, Apr 2, 2015).

Interestingly, more money was invested in renewables in 2011, around $278.8 billion (Mooney, March 31, 2015). Because capital costs have fallen so much since then, every billion committed in 2014 produced many more megawatts of capacity, meaning 2014 is still the largest overall investment in renewable energy (Mooney, March 31, 2015). The report estimates that this new investment means 9.1% of total world energy generation is now coming from renewables, which is equivalent to 1.3 gigatonnes less carbon in the atmosphere (Mooney, March 31, 2015). Even so, at this rate, renewables would make up only 20% of world energy production in 2030, and there is very little chance atmosphereic levels of carbon dioxide would stay below 450 parts per million, the scientifically identified safe amount for Earth (Mooney, March 31, 2015). Although this report indicates that the renewable industry is growing quickly and steadily, much more needs to be done in government to create policy incentives and speed up the process of renewable investment.

Hruska, Joel. Gobal Investment in Renewable Energy Skyrockets. April 2, 2015. [http://www.extremetech.com/extreme/202579-global-investment-in-renewable-energy-skyrockets]

Mooney, Chris. Renewable energy is growing very, very fast. Its just still not fast enough. March 31, 2015. [http://www.washingtonpost.com/news/energy-environment/wp/2015/03/31/renewable-energy-is-growing-very-very-fast-its-just-still-not-fast-enough/]

Colorado Utilities Deploy Renewable Energy Because of Lower Costs

by Liza Farr

Amidst the many states scrambling to meet new emission reduction requirements mandated by both state and federal government, Colorado stands out as a problem child turning a new leaf. Colorado gets 58% of its energy from coal-fired powered plants, but two utilities in particular have announced plans recently for new renewable energy sources (Jaffe, April 12, 2015). Rather than citing the reduction requirements as the reason for this shift, spokesmen for the companies say the falling costs of renewable energy sources has allowed them to add on these projects (Jaffe, April 12, 2015). Back in 2012, the renewable energy market was dimming, despite the state requirements that had just been passed (Jaffe, January 22, 2012). However, loss of state incentives and uncertainty about the future of federal subsidies were causing renewable energy growth to slow (Jaffe, January 22, 2012). Experts in 2012 predicted that the biggest boost for renewables would come as they became cheaper and more competitive with conventional energy sources, and the present day reveals these predictions to be true (Jaffe, January 22, 2015).

Platte River Power is adding 22 MW of solar photovoltaics near Wellington, Colorado, and Tri-State Generation and Transmission Association is adding a 150 MW wind farm in Kit Carson County to their energy sources, a 25-year contract costing $240 million (Jaffe, April 12, 2015). As of now, Platte River Power gets less than 5% of its power from wind, none from solar, and 20% from hydropower (Jaffe, April 12, 2015). Tristate gets a similar figure of 24% from wind and hydropower, and the company has added 800 MW of renewable power since 2008 (Jaffe, April 12, 2015). These two companies are both wholesale electricity generators and do not fall under the standards for renewable energy by the state requirement that utilities get 30% of their electricity from renewables by 2020. Still, the companies have begun shifting away from coal-fired plants, at least partly because of the cost competitiveness of renewables (Jaffe, April 12, 2015). The companies cite transmission lines as the largest challenge: Tri-State is building a 72-mile line between Burlington and Wray that costs $40 million. The state could encourage more utilities to utilize the renewable energy resources if they pursued a project similar to the Texas Competitive Renewable Energy Zone, through which the state built 3,600 miles of transmission lines to connect Western Texas wind power to major urban areas (Ross, March 2015). By building the infrastructure with State money, the government can cut out the costly and time consuming aspect of renewable energy projects for utilities.

The EPA recently passed regulations dictating that carbon dioxide emissions be cut by 30% nationwide by 2030 (Jaffe, July 29, 2014). Targets are set for each state based on local power supply and existing energy policies, and Colorado was assigned a 35% reduction target (Jaffe, July 29, 2014). Colorado environmental policy groups commented that the policies already in place make this a very doable target Colorado. In fact, simply the existing programs would cut state emissions by 28% by 2030 (Jaffe, July 28, 2014). Despite this, a spokesman for Tri-State stated that "EPA is placing state regulators in a difficult position to craft complex plans in a short period of time" (Jaffe, July 28, 2014). The Colorado Mining Association also unsurprisingly commented that the regulations would hurt jobs and the economy, estimating 9,000 mining jobs lost (Jaffe, July 28, 2014). Some pushback against regulations is to be expected, but the comment by Tri-State spokesmen indicates that the company is pursuing renewables only because of their cost effectiveness, and would prefer to pursue them at their own pace, rather than a mandated pace that may hurt profits.

Renewables do have some ground to cover before becoming a truly cheaper option compared with coal, since the former requires infrastructure and solar farms still to be built, unlike coal, which is already being turned into energy in existing power plants. Platte River Power, on the other hand, is already well on its way to raising its renewable energy generation to 32% by 2016, although this includes large hydropower, and renewable energy credits (Jaffe, April 12, 2015). They also noted that their goal of cutting carbon by 80% by 2050 must be kept in the context of maintaining reliability and keep rates low among wholesale suppliers, as they currently charge only 5.5 cents per kilowatt hour to municipal systems (Jaffe, April 12, 2015). Government help in paying for infrastructure or providing incentives to make building renewable energy cost competitive with existing coal-fired power plants is necessary to spur on renewable development in Colorado, as well as the rest of the nation.

Jaffe, Mark. Colorado's Future in Renewable Energy Dims after Years of Growth. January 22, 2012. [http://www.denverpost.com/ci_19789027?source=infinite].
Jaffe, Mark. EPA Greenhouse-Gas Plan could up Pressure on Colorado Power Utilities. July 28, 2014. [http://www.denverpost.com/business/ci_26232488/epa-greenhouse-gas-plan-could-up-pressure-colorado].
Jaffe, Mark. Colorado's big coal-burning utilities take a turn to renewable energy. April 12, 2015. [http://www.denverpost.com/business/ci_27890920/colorados-big-coal-burning-utilities-take-turn-renewable].
Ross, Dale. Mayor: Why My Texas Town Ditched Fossil Fuel. March 27, 2015. [http://time.com/3761952/georgetown-texas-fossil-fuel-renewable-energy/]

Potential $8-Billion Green Energy Initiative for Los Angeles

by Alexander Flores

As of September 2014, an $8-billion dollar green energy project has been proposed to link one of the nation's largest wind farms to one of the world's biggest energy storage facilities. This first initiative of its kind in the United States was strung together by four companies: Duke-American Transmission Co., Dresser-Rand, Magnum Energy, and Pathfinder Renewable Wind Energy, in hopes to provide large quantities of clean electricity to the Los Angeles area by 2023. This project in particular would involve the construction of one of the largest wind farms in Wyoming and one of the largest energy storage facilities in Utah with a 525-mile electric transmission line connecting the two sites. The $4-billion dollar wind farm would be built, owned, and operated by Pathfinder Renewable Wind Energy and would generate 2,100 megawatts of electricity. Along with this, the $1.5-billion energy storage facility would be installed in Delta, Utah, by Pathfinder Renewable Wind Energy, Magnum Energy, and Dresser-Rand. This "compressed air energy storage" would be composed of four vertical caverns that would be one-quarter mile high, 290 feet in diameter, and 41 million cubic feet in volume. As a unit, the four caverns combined are proposed to store 60,000 megawatt-hours of electricity. The $2.6-billion, 525-mile electric transmission line would be built by Duke-American Transmission in order to transport electricity from the Wyoming wind farm to the energy storage facility in Utah. An existing 490-mile transmission line would also aid in electricity transport from Utah to the Los Angeles area. The underground storage facility would help solve the intermittency of renewable energy. Without wind, wind farms can't produce electricity and without sunlight, solar farms can't produce electricity. With the ideal utilization of the electric transmission line, the wind farm would be able to function similarly to a natural gas power plant, which is capable of delivering large amounts of electricity whenever necessary based on demand. For instance, when customer demand is low, the storage facility would use electricity from the wind farm to compress and inject high-pressure air into the energy storage caverns. Then, when customer demand is high, the stored, high-pressure compressed air would be combined with minimal natural gas to power eight generators for electricity production.

This energy storage facility would ultimately reduce the need for expensive backup power plants and power lines when there isn't any wind or sun (at night) to be utilized by a traditional wind or solar farm to produce electricity. If these four companies were to have their formal proposal approved by the Southern California Public Power Authority later in 2015, this project could potentially generate 9.2 million megawatt-hours per year and serve approximately 1.2 million Los Angeles-area homes once in full effect. For our sake, in terms of a "greener" future, let's all hope this one goes the distance.

Downey, John. 2014. Duke Energy joint venture part of $8 billion bid to supply green energy to Southern California. Charlotte Business Journal. Sept. 23, 2014. (http://www.bizjournals.com/charlotte/blog/energy/2014/09/duke-energy-jv-takes-part-in-8-billion-bid-to.html)

Duke American Transmission Co. 2015. (http://www.datcllc.com/news-releases/8-billion-green-energy-initiative-proposed-for-los-angeles/)

Funding for Energy Initiatives in Africa: US-Africa Clean Energy Finance Initiative

by Alison Kibe

The US-Africa Clean Energy Finance (ACEF) initiative launched at the UN Conference on Sustainable Development in 2012. As of August 2014, the US had pledged $30 million to fund ACEF. The United States Trade and Development Agency's (USTDA) January 2015 press release announced the two entities in charge of funding AFEC, the Overseas Private Investment Corporation (OPIC) and USTDA, have both obtained initial funds for AFEC projects. Both organizations are involved with connecting private American businesses to international development projects. The goal of ACEF is to promote privately financed clean energy projects with the hope that ACEF acts as a catalyst for economic development and promotes US foreign policy goals in Africa.

According to Sarah Carta, the AFEC Program Manager, ACEF funds will be used to provide countries, includes Ethiopia, Tanzania, and Rwanda, with greater access to electricity. PennEnergy reports that President Obama's Power Africa Initiative used ACEF to partially fund its first completed project—a $23.7 million 8.5 MW solar field in the Agahozo-Shalom Youth Village. The solar field will not only bring electricity to a village that cares for children orphaned during the Rwandan genocide, but the rent paid to lease the land where the field is located will help pay for expenses related to the village's charity. The USTDA is using ACEF money to invest in solar corporations like Off-Grid Electric that employs local workers to maintain solar energy home kits in Tanzania. Funds are also being put into African research and development companies.

AFEC is only one among many groups interested in investing in energy projects. As venture capitalists become more prominent in the US, the venture capitalist market is heating up in African countries as well. Tom Jackson, writing for an African business and investment website, explains that venture capitalists are especially focused on energy initiatives as many African countries continue to "leap-frog certain technologies." The outcomes and final impacts on economic development remain to be seen.

Carta, Sarah. "Highlights from the Field: ACEF supports African renewable energy projects for enduring development." The OPIC Blog. August 12, 2014. http://www.opic.gov/blog/opic-in-action/highlights-from-the-field-acef-supports-african-renewable-energy-projects-for-enduring-development

PennEnergy "East Africa's first utility-scale solar field to officially open." PennWell Corporation. February 2, 2015.
http://www.pennenergy.com/articles/pennenergy/2015/01/east-africa-s-first-utility-scale-solar-power-field-officially-open.html

Jackson, Tom. "Africa Tech Trends: Cash flowing in for start-ups; Uber's Kenyan safari." How We Made It In Africa. February 2, 2015.
http://www.howwemadeitinafrica.com/africa-tech-trends-cash-flowing-in-for-start-ups-ubers-kenyan-safari/46570/

United States Trade and Development Agency. Africa-Focused Renewable Energy Initiative Reaches Full Commitment of Initial $20 Million Investment. [Press release] January 12, 2015.
http://www.ustda.gov/news/pressreleases/2015/SubSaharanAfrica/OPIC/AfricaFocusedRenewableEnergyInitiative.asp

Crowdsourcing to Eliminate Energy Poverty
by Alison Kibe

Energy costs make up a disproportionate amount of a low income household spending so energy prices, recently at 3.2% increase per year, create a greater burden on lower incomes (Hodge, 2014). As a result, some households cannot afford to heat or cool their homes without assistance. Under some circumstances, households may even forego energy use in months they need it most. This is not a small issue, with the official poverty rate in 2013 at 14.5%—or about 45.3 million people (US Census Bureau, 2014). Gridmates, a startup founded in 2014, thought of a new platform to reach those in need—through the electricity grid.

According to the Gridmates website (Gridmates.com), Gridmates uses a software as a service platform operated through utilities to form a network between utilities customers, corporations, nonprofits, and those in need. Customers and corporations can pay for and "send" energy through the Gridmates website. Although an aim of the service is to help those in need, users could send energy to family members, friends, and nonprofit organizations. Gridmates also allows any customers who generate electricity at home to choose if and how they want to share surplus electricity with others.

Gridmates is currently using their platform to crowd source energy funding for the Community First! Village. The project, run by the City of Austin and Mobile Loaves and Fishes, is based in Southeast Austin and will provide affordable and sustainable housing to 240 disabled and chronically homeless individuals. Community First! Village's first year total electricity costs are estimated at $213,000, or $2.50 per person per day (Tull, 2015). As of March 2015, Gridmates had raised $85,000 (Gridmates.com).

This type of service is what the National Institute of Standards and Technology expects to see in the future of energy grids, or what is currently being called Grid 3.0 (Greer, 2015; NIST, 2014). Grid 3.0 is concerned with leveraging current Grid 2.0 technology (think smart meters) to change how customers and utilities interact with energy grids. With over 43 million US households having smart meters as of 2012, other services using platforms that rely on data available through smart grids could become more widely available (US Energy Information Administration, 2014).

Greer, Chris. "Legacy Grid, Smart Grid." Energy Biz Magazine, Winter 2015 Issue. http://www.energybiz.com/magazine/article/392097/legacy-grid-smart-grid

Gridmates.com http://www.gridmates.com/

Hodge, T., 2014. Residential electricity prices are rising. US Energy Information Administration. http://www.eia.gov/todayinenergy/detail.cfm?id=17791

Tull, S., 2015. Fact Sheet: Energy Sharing Campaign: Benefiting Community First! Village And Creating The World's First Community Powered By Crowdsourced Energy. Gridmates.

National Institute of Standards and Technology, 2014. NIST Framework and Roadmap for Smart Grid Interoperability Standards, Release 3.0. http://dx.doi.org/10.6028/NIST.SP.1108r3

US Census Bureau, 2014. Poverty: 2013 Highlights. http://www.census.gov/hhes/www/poverty/about/overview/

US Energy Information Administration, 2014. How many smart meters are installed in the US and who has them? http://www.eia.gov/tools/faqs/faq.cfm?id=108&t=3

Changes in Clean Energy Investment in 2014: End of Year Recap

by Melanie Paty

On January 9th, 2015, Bloomberg New Energy Finance submitted a press release on the strong performance of clean energy investments in 2014. The overall investment in clean energy reached $310 billion, a 16% increase from 2013, but 2011 still holds the record at $317 billion. However, it was the biggest increase of new investment in clean energy since 2011. Government funded research and development increased 14% and corporate increased 15%. Private equity and venture capital investments increased 16%, but overall investment is still three times below 2008 levels. In terms of region, the most investment came from the United States, China, and Europe. European investment increased only 1% since 2013, but is still the highest at $66 billion. China increased a whopping 32% to $89.5 billion. Clean energy investment in the United States experienced a smaller increase of only 8% reaching $51.8, $15.5 billion of which went to utility scale asset finance. US investment in solar increased by 39% whereas investment in wind decreased by more than 50%. India and Brazil both reached $7.9 billion in clean energy investments, an 88% increase for the former and a 14% increase for the latter. French investment increased by 26% due to the installation of Europe's largest solar PV plant with 300MW capacity.

The majority of investment dollars went to asset finance of renewable energy including at least seven $1 billion offshore wind projects in Europe and large-scale solar projects in Japan, South Africa, Kenya, and Ontario. The second largest investment category was small-distributed capacity projects of less than 1MW, which primarily took the form of rooftop solar installations. Approximately 50% of the investments were in solar, marking its highest share to date. Investment in wind increased by 11% and investment in smart energy technologies like smart grid and storage were the third largest category. Finally, green bonds reached a record high of $38 billion, 2.5 times more than 2013 investment. It will be interesting to see how clean energy fares in 2015 as it is likely the effects of the recent drop in oil price have yet to be felt.

Bloomberg 1 (http://about.bnef.com/press-releases/rebound-clean-energy-investment-2014-beats-expectations/).

Bloomberg 2 (http://about.bnef.com/presentations/clean-energy-investment-q4-2014-fact-pack/).

Global Consequences of Crude Oil Price Reductions
by Ali Siddiqui

The current global surplus and price reduction for crude oil has affected the ability of countries to meet renewable energy goals. Indonesia's new government recently passed legislation approving a subsidy to biodiesel production in an effort to reach their predicted consumption goal of 1.7 million kiloliters in 2015. However, this legislation did not take into account the current prices of crude oil. The lower prices of crude oil have cut both domestic and international demand for biodiesel. Biodiesel has become the more expensive alternative and therefore the forecasted consumption target used to help shape the government sponsored subsidy program may now be set too high.

Biodiesel is a mixture or blend of diesel derived from crude oil and biofuel mixed at a certain ratio. Since 2013, Indonesia blends the two together at 10% and in 2014 commanded 20% blending from power plants. Biodiesel is a leading driver of palm oil prices as well. Indonesia had sought to increase palm oil prices if their policies had taken place, however, now analysts are forecasting that demand for biodiesel may only reach 1 million kiloliters in 2015. The lower price of crude oil cutting demand has forced policy makers to rethink a formula in order to ensure biodiesel price is competitive. This formula will include biofuel production costs plus the price of crude palm oil prices as opposed to the previously used formula, which utilized the Mean of Platts Singapore diesel price plus "3.48% as reference in tenders".

Not everyone in Indonesia is skeptical about reaching the target forecasted first. One analyst from Mandiri Sekuritas, a capital market service provider, believes that the subsidy will increase demand for palm oil by 900,000 tons. Bangun estimates an increase in palm oil production by 1.5 million tons since their reported 29.5 million tons in 2014. Yet, one must still consider the potential and real impacts of lower crude oil prices. The secretary general of the Biofuel Producers Association stated "Biodiesel is uneconomical at the moment", so the economic confidence about biodiesel and this recently passed legislation is still ambivalent.

Rusmana, Yoga., Listiyorini, Eko. Crude Oil's collapse seen causing Indonesia to miss Biofuel goal. March 2nd 2015. http://www.bloomberg.com/news/articles/2015-03-01/crude-oil-collapse-seen-causing-indonesia-to-miss-biofuel-target

Are Energy Credits Enough?
by Ali Siddiqui

Last year in Pennsylvania, the price of solar energy credits began to decrease. Energy credits or renewable energy credits are government sponsored economic incentives for companies to use more alternative sources of energy. The credits in Pennsylvania work by having each megawatt hour of power produced by alternative energy qualify the company producing that energy for a renewable energy certificate, which can be sold or traded to other companies. As the solar energy credit prices fell, other alternative energy credits saw a rise, which caused financial analysts to forecast more investment into alternative energy.

Last year geothermal, wind, and landfill gas credits were being valued on average around $10 and this year their prices are doubled. The reasoning behind this doubling, besides the general movement towards green energy, comes from the projection by the Pennsylvania Public Utilities Commission that demand for these particular alternative energy credits will be much greater than their supply.

Companies such as the Milton Regional Sewer Authority in Northumberland County have taken these projections to heart and have acted by expanding alternative energy production through creating a larger wastewater plant to generate 2 megawatts per day of electricity from the use of biogas.

Renewable energy credits bought by companies are interesting because they are also not propriety of a single company and can be traded, which can impact a companies finances on a small scale. Now, although there is heavy optimism towards these energy credits, recent years have shown that the value of these solar credits tends to be volatile. Tier 1 resource credits, credits given to energy produced not by solar photovoltaic cells, however, have seen a steady rise now reaching 105% of their original value since 2010. This ambivalence about energy credits stems from the volatility of the credit market and their small impact on the companies' finances. To improve this market, some critics argue that legislation is what needs to be changed in order to make energy credit incentives more popular.

Litvak, Anya. Renewable energy credits rose last year, but is it enough. March 3rd 2015. http://powersource.post-gazette.com/powersource/policy-powersource/2015/03/02/Groups-highlight-risk-of-crude-oil-train-derailments-in-Pittsburgh/stories/201503020128

About the Authors

The authors of this book are students at the Claremont Colleges. The book is a work product of Biology 137: Environment, Economics, and Policy Clinic taught by Emil Morhardt in the W.M. Keck Science Department of Claremont McKenna, Pitzer, and Scripps Colleges.

The students' task was to write journalistic summaries of descriptions of interesting new innovations in energy culled from whatever sources they wished, capturing the essence of the innovations but eschewing technical terms to the extent possible—to become, in effect, science writers. The summaries were due weekly and were returned with editorial comments shortly thereafter.

The editor is Roberts Professor of Environmental Biology at Claremont McKenna, Pitzer, and Scripps colleges. He remembers how difficult it is to learn to write and appreciates the professionalism shown by these students.

Index

"Ariel" suspension light 30
Abengoa................................. 157
Achates Power.................. 208, 209
Acoustic Building Infiltration
    Measurement System 49, 50
ADSP Consulting 126
Advanced Grid Innovation
    Laboratory for Energy 95
Advanced Microgrid Solutions ... 68,
    69
Afsluitdijk causeway 104
Airborne Wind Energy (AWE).... 179
AllCell Technologies LLC 62
AllEarth Renewables 127, 128
AltaRock Energy, Inc............... 187
Amazon............... 50, 51, 278, 289
American University................. 246
Apple 24, 25, 62, 105, 135, 136,
    277, 278, 289, 290
Ar Vag Tredan 82
Arctic National Wildlife Refuge
    (ANWR) 298
Argonne National Laboratory...... 49
Atlas Energy Systems....... 192, 193
Barber, Mariah Valerie.. 23, 24, 25,
    26, 62, 95, 103, 104, 117, 153,
    159, 215, 235
Barnett Shale........... 245, 246, 256
BaseTrace 246
battery, redox flow 92
Bayer Material Science............. 202
Bell Labs.............................. 111
BigApps NYC........................... 71
Bill & Melinda Gates Foundation
    230
BioCarbon Engineering 204
biomimetics 119
Bloomberg ...64, 65, 136, 152, 187,
    188, 201, 202, 236, 259, 278,
    280, 281, 288, 307, 308, 309,
    310, 322, 323
Bluegas.................................. 232
Boeing 149, 202
BrightSource Energy........ 153, 154

Bristol BioEnergy Centre.......... 230
Bristol Robotics Laboratory 230
Brown, Hannah. 27, 29, 30, 31, 63,
    118, 120, 121, 122, 199, 200,
    257, 305
Bundogji, Nour64, 65, 66, 123, 125,
    126, 160, 202, 236, 258, 261,
    262, 306
butanol.................................. 221
California Institute of Technology
    128
California State University,
    Northridge........................... 84
Caltech 128, 129, 130
Capper, Jessie..... 32, 34, 127, 128,
    175, 203, 204, 265, 266, 267,
    308, 309, 310
CarbFix....................... 257, 258
Carnegie Melon University........ 114
Carnegie Wave Energy............. 161
Center for Research on
    Environmental Decisions (CRED)
    269
China National Renewable Energy
    Center,............................... 213
China University of Geosciences
    108
China-Kenya Solid State Lighting
    Technology Transfer Center.. 117
Cisco Systems................... 29, 120
Citi 306, 307
Clean Current Power Systems . 159,
    160
Clean Energy Collective............ 148
CleanTechnica 128, 130, 310
Clemson University 66, 67, 68
Columbia Engineering............. 126
Community Solar for Community
    Action 309
Cool Choices 59
Coolerado............................ 23, 24
Copenhagen..29, 30, 120, 121, 286
CorPower Ocean...................... 168
Crescent Dunes 154, 155, 156

CubeSat 83
Décision 202
Delos-Reyes Morrow Pressure
    Device 171
DeltaStream technology............172
desalination.............. 122, 123, 162
Desert Sunlight Solar Farm 145,
    146
Detroit Zoo 216, 217
Dongo Kundu and Liquid Natural
    Gas project...........................176
Dresser-Rand 318
Drexel 63, 64, 74, 75
Drexel University.................63, 74
Duke Energy Renewables 99
Duke-American Transmission Co.
   318
Dynamic Glass 39
Dyson...................................30, 31
Eagle Ford Shale 251, 252
Earthtronics...............................43
Edison Electric Institute ... 151, 312
Elder, Alex..... 35, 36, 68, 105, 130,
    131, 176, 178, 216, 231, 269,
    311
Electric Power Research Institute92
electrification.............................60
electrodialysis..........................122
Emil Morhardt, J. Emil...............21
Endesa.....................................211
Energous.........................106, 107
Energy Bridge............................32
Energy Efficiency & Renewable
    Energy.............................49, 50
Energy Star 42, 43, 275, 276
Energy Star Home Advisor..........42
Energy Storage Association, The .91
Energy Vulture.com..................138
EnerVault...................................89
Enhanced Geothermal Systems
    (EGS)...................................188
Essess...................................57, 58
ethylene 52
eWind Solutions 179, 180
Exelon...................... 98, 196, 197
exoelectrogens.........................110
Farr, Liza. 132, 133, 154, 206, 207,
    237, 270, 272, 273, 312, 313,
    315, 316
FirstFuel 33, 57
Flores, Alexander37, 38, 39, 40, 69,
    134, 178, 208, 217, 232, 233,
    238, 318
Florida Center for Renewable
    Chemicals and Fuels 237
Florida State University 76

flywheel 80, 82, 92
flywheels 61, 76, 79, 80, 92, 93,
    209
Forbes 56, 263, 264, 283, 296, 307
FracEnsure 246
Fundy Tidal Incorporated 160
gadolinia 67
gamification...............................58
Gasa..................................276, 277
GENeco 206, 219
Georgia Institute of Technology.140
Georgia Tech Research Corp.....172
Gevo, Inc. 222
Goodman, Dylan41, 42, 43, 44, 45,
    46, 47, 135, 136, 161, 191, 218,
    275
Google 27, 32, 62, 63, 105, 153,
    154, 178, 278, 289, 290, 306,
    313, 314, 315
graphene 72, 73, 83, 108, 138, 139
GreatPoint Energy 232, 233
GreenTechMedia........................33
Gresham, Oregon 34, 35
Gridmates 321, 322
GTM Media................................99
Harbourton Alternative Energy .166
Highlands EnviroFuels 222
hohlarum 194
Honeywell.................................32
HVAC 50, 55, 57
Hybrid Social Solutions 132, 133
Hydrogen and Fuel Cell
    SUPERGEN Hub...........212, 241
Ice Bear....................................69
Ice Energy 69, 70
Iceland 185, 186, 257, 258
iGrenEnergi......................123, 124
Imperial College.......................225
InEntec 233
International Renewable Energy
    Agency (IRENA).....................172
International Thermonuclear
    Experimental Reactor (ITER) .195
isobutanol 222
Ivanpah 153, 154, 155, 156
Jain Irrigation Systems.............122
Jet Propulsion Laboratory...84, 129
Kibe, Alison 47, 49, 70, 71, 137,
    138, 139, 140, 179, 192, 276,
    320, 321
Kocaeli University......................84
Kyocera TCL Solar 182
La Rance Power Plant 172
Lake Turkana Wind Power Project
   176

Lamont Doherty Earth Observatory 203
Laser Power Systems (LPS) 191
Lawrence Berkeley National Laboratory 62, 312, 313
Lawrence Livermore National Laboratory 194, 195
LED . 29, 30, 31, 37, 38, 43, 44, 65, 112, 120, 306, 308
Lee, Briton ... 50, 51, 52, 72, 74, 75, 96, 106, 107, 141, 219, 277
LiDAR 57
LightSail 89
Liquid Air Energy Storage 91
LiquiGlide 35, 36, 44
Lucid Energy 166
Lucideon 36, 37
Lumentech 262
M3 Wave Project 171
Magnum Energy 318
Makani Power 178
Marcellus Shale 252
Massachusetts Institute of Technology 35, 113, 258
MeyGen 167
MiaSolé 134, 135
Michigan State University 125, 200, 266
Microsoft 33, 278
Mirai 64
MIT Energy Initiative 44, 45
M-KOPA Solar 139, 140
Mojave Desert 21, 142, 153, 154, 157, 285
Morhardt, J. Emil 4, 5, 6, 53, 76, 77, 78, 79, 80, 81, 82, 83, 84, 85, 86, 108, 109, 110, 111, 142, 144, 162, 163, 164, 180, 181, 209, 220, 221, 243, 245, 246, 247, 248, 249, 250, 251, 252, 253, 254, 255, 256, 327
M-PESA 140
Mutriku wave project 168
MXene 63
MyEnergy 28
Nagar, Niti ... 54, 87, 111, 112, 157, 165, 166, 193, 223, 224, 225, 226
nanomaterials 74, 138
NASA 129, 130, 179, 180
National Appliance Energy Conservation Act 46
National Grid 70, 71, 101
National Institute of Building Sciences 50

National Renewable Energy Laboratory 40, 41
National Space Society 129, 130
National Synchrotron Light Source 87
Nest 27, 28
New Ventures 124
New Visual Media Group 36
Nicaragua 186, 187
Nissan 77, 207, 210, 211
Nissan Leaf 77, 207
NordLink 100
Northrop Grumman . 128, 129, 130
Northumbria University 78
O'Neill, Shannon . 88, 97, 114, 145, 146, 147, 167, 185, 186, 227, 228, 278
Oak Ridge National Library 25, 103
Olkaria Geothermal facility 265
Olkaria project 266
OPOWER 32, 33, 34
organic photovoltaic (OPV) cells 141
ORNL 25, 26, 54, 55, 103, 104
Overseas Private Investment Corporation (OPIC) 320
Oxfam 230
Pacific Northwest National Laboratory 189
pallid sturgeon 169, 170
Pathfinder Renewable Wind Energy 318
Patronicity 216, 217
Paty, Melanie 89, 148, 168, 210, 279, 280, 281, 282, 284, 285, 286, 287, 322
Pennsylvania State University 58
perovskite 136, 137
Picasolar Inc. 140
Plasma Enhanced Melter (PEM) 233
Pluvia 166
polyallylamine 189, 190
polyethylene 52, 53, 222
PosiGen 121, 122
POWER (Partners Offering a Water Energy Revolution) 172
Powerley 31, 32
PowerScan 32
propane ... 108, 109, 206, 225, 226, 254
Purdue University ... 105, 106, 107, 108
Rawlemon 150, 151
Redman, Chad 90, 91, 92, 93, 149, 150, 195, 212, 240
REDstack BV 104
Reforming the Energy Vision 95

Regin Paradise Consulting........124
Rice University.................218, 219
Roosevelt Island Tidal Energy
    Project (RITE)......................171
RREAL 309, 310
Sabien Technology...............26, 27
sage grouse 296, 297
Sea Shepherd Conservation Society
   203
Seageneration...........................167
Semitrex.............................56, 57
Sensity Systems 29, 120
Siddiqui, Ali99, 187, 213, 229, 288,
    289, 290, 291, 292, 323, 324
Siftung Solarenergie Foundation
   132
Siluria Technologies 231, 232
Skystream 178, 179
Skyview monitoring software 179
Solana.............................155, 157
Solar Cloth Company........133, 134
Solar Impulse 202, 203
Solar Vets program...........278, 279
SolarCity 148, 281
SolarReserve....................155, 156
SolarWindow 47
Solazyme, Inc. .. 215, 216, 217, 218
SolePower................................114
Southern California Edison Co. .68,
    69
Space Solar Initiative........128, 130
Spectrolab 149, 150
Stanford Energy System
    Innovations 41
Stanford University 41, 54, 75, 108,
    109, 136
Staples Inc. 45
Star Renewable Energy.............115
storage, pumped hydroelectric....93
StoreDot.............................71, 72
SunEdison . 89, 131, 132, 274, 280
SunPower.. 41, 127, 128, 135, 136,
    311
SunRun...................................148
SunShare 149
Sunyale Africa Limited.............118
supercapacitor .. 61, 64, 79, 80, 82,
    83, 84, 85, 86, 112, 209
syngas.............................231, 233
Tendril 38, 39
Tendril Energy Services
    Management...........................38
Tesla Motors Inc., ...62, 64
thorium...........................191, 192
Tidal Energy Ltd. (TEL) 172

Tidal Lagoon Power.. 160, 161, 170,
    171
Toshiba 235, 236
Transatomic Power...........192, 193
transparent luminescent solar
    concentrator 125
TREASORES................................65
triboelectric nanogenerators 111
Tumalow 70, 71
United States Trade and
    Development Agency (USTDA)
   320
University in Suwon, South Korea
   137
University of Arkansas......140, 141
University of Colorado Boulder ..52,
    53
University of Florida 237
University of Georgia 226
University of Manchester Institute
    of Biotechnology (MIB) 225
University of Michigan 223, 224
University of Strathclyde...........129
University of Texas .. 224, 245, 247,
    248, 251, 252, 256
University of Texas at El Paso ...224
University of Turku...................225
University of Twente 104, 105
University of West England (UWE
    Bristol) 230
University of Wisconsin-Madison87,
    88
urine-tricity.............................230
US Department of Energy25, 26, 71,
    103, 104
US Green Building Council.........48
US Naval Academy......................74
US-Africa Clean Energy Finance
    (ACEF)..................................320
Vanderbilt University.................86
Vectorform 31, 32
Vectren....................................28
Verge, The 64, 65
View Inc. 39
Virginia Tech 236, 237, 241
Wang, Abigail 57, 58, 59, 100, 151,
    171, 230, 241, 296, 298, 299,
    301, 302
WattUp..............................106, 107
Whole Building Design Guide 50
WindFloat technology 182
Wisconsin Energy Institute.........87
Woods Hole Oceanographic
    Institution 203
World Resources Institute.........229
Xzeres 178, 179

ZigBee.. 31